The Limits of Life Writing

In the age of social media, life writing is ubiquitous. But if life writing is now almost universal—engaged with on our phones; reported in our news; the generator of capital, no less—then what are the limits of life writing? Where does it begin and end? Do we live in a culture of life writing that has no limits? Life writing—as both a practice and a scholarly discipline—is itself markedly concerned with limits: the limits of literature, of genres, of history, of social protocols, of personal experience and forms of identity, and of memory.

By attending to limits, border cases, hybridity, generic complexities, formal ambiguities, and extra-literary expressions of life writing, *The Limits of Life Writing* offers new insights into the nature of auto/biographical writing in contemporary culture. The contributions to this book deal with subjects and forms of life writing that test the limits of identity and the tradition of life writing. The liminal case studies explored include magical-realist fiction, graphic memoir, confessional poetry, and personal blogs. They also explore the ethical limits of representation found in Holocaust life writing, the importance of ficto-critical memoir as a form of resistance for trans writers, and the use of "postmemoir" to navigate the traumas of diasporic experience. In addition, *The Limits of Life Writing* goes beyond the conventional limits of life-writing scholarship to consider how writers themselves experience limits in the creation of life writing, offering a work of life writing that is itself concerned with charting the limits of auto/biographical expression.

The chapters in this book were originally published as a special issue of *Life Writing*.

David McCooey is an award-winning poet, critic, and editor. He is the author of *Artful Histories: Modern Australian Autobiography* (1996/2009), and he has published widely on life writing and poetry. He is the Deputy General Editor of *The Literature of Australia* (2009), and his most recent book of poems is *Star Struck* (2016). He is Professor in Writing and Literature at Deakin University in Victoria, Australia. His website is davidmccooey.com.

Maria Takolander is a poet, fiction writer, and scholar. Her most recent book of poems is *The End of the World* (Giramondo 2014). A widely published scholar of magical realism, she has also published numerous essays theorising creativity in relation to mental illness, embodied cognition, and sociomaterial practice. She is an Associate Professor in Writing and Literature at Deakin University in Victoria, Australia. Her website is mariatakolander.com.

The Limits of Life Writing

Edited by
David McCooey and Maria Takolander

LONDON AND NEW YORK

First published 2018
by Routledge
2 Park Square, Milton Park, Abingdon, Oxon, OX14 4RN, UK

and by Routledge
711 Third Avenue, New York, NY 10017, USA

Routledge is an imprint of the Taylor & Francis Group, an informa business

British Library Cataloguing in Publication Data
A catalogue record for this book is available from the British Library

ISBN13: 978-0-8153-9191-3

Typeset in MinionPro
by diacriTech, Chennai

Publisher's Note
The publisher accepts responsibility for any inconsistencies that may have arisen during the conversion of this book from journal articles to book chapters, namely the possible inclusion of journal terminology.

Contents

Citation Information

The chapters in this book were originally published in *Life Writing*, volume 14, issue 3 (September 2017). When citing this material, please use the original page numbering for each article, as follows:

Introduction
The Limits of Life Writing
David McCooey
Life Writing, volume 14, issue 3 (September 2017) pp. 277–280

Chapter 1
Joe Sacco's Australian Story
Gillian Whitlock
Life Writing, volume 14, issue 3 (September 2017) pp. 283–295

Chapter 2
Malala Yousafzai, Life Narrative and the Collaborative Archive
Kate Douglas
Life Writing, volume 14, issue 3 (September 2017) pp. 297–311

Chapter 3
Remembering Violence in Alice Pung's Her Father's Daughter: The Postmemoir and Diasporisation
Anne Brewster
Life Writing, volume 14, issue 3 (September 2017) pp. 313–325

Chapter 4
Witnessing Moral Compromise: 'Privilege', Judgement and Holocaust Testimony
Adam Brown
Life Writing, volume 14, issue 3 (September 2017) pp. 327–339

Chapter 5
'A Thing May Happen and be a Total Lie': Artifice and Trauma in Tim O'Brien's Magical Realist Life Writing
Jo Langdon
Life Writing, volume 14, issue 3 (September 2017) pp. 341–355

Chapter 6
Forms of Resistance: Uses of Memoir, Theory, and Fiction in Trans Life Writing
Juliet Jacques
Life Writing, volume 14, issue 3 (September 2017) pp. 357–370

Chapter 7
Confessional Poetry and the Materialisation of an Autobiographical Self
Maria Takolander
Life Writing, volume 14, issue 3 (September 2017) pp. 371–383

Reflection
I Guess What You Say is True
Oliver Driscoll
Life Writing, volume 14, issue 3 (September 2017) pp. 387–400

Notes on Contributors

Anne Brewster is an Associate Professor at the University of New South Wales, Australia. Her books include *Reading Aboriginal Women's Life Writing* (1995/2016) and *Giving this Country a Memory: Contemporary Aboriginal Voices of Australia* (2016). She is currently finishing a book with Sue Kossew titled *Rethinking the Victim: Violence, Gender and Australian Women's Writing* (Routledge).

Adam Brown is a senior lecturer in digital media at Deakin University, Australia. You can find out more about Adam and his work at adamgbrown.wordpress.com.

Kate Douglas is a Professor in the College of Humanities, Arts, and Social Sciences at Flinders University, Australia. She is the author of *Contesting Childhood: Autobiography, Trauma, and Memory* (Rutgers 2010) and the co-author (with Anna Poletti) of *Life Narratives and Youth Culture: Representation, Agency and Participation* (Palgrave 2016).

Oliver Driscoll is a PhD candidate at Deakin University, Australia. He co-runs the Slow Canoe Live Journal in Melbourne, Victoria, and manages fiction submissions at the Australian journal *Overland*.

Juliet Jacques is a writer and filmmaker based in London. She has published two books, most recently *Trans: A Memoir* (Verso 2015). Her short fiction, essays, and journalism have appeared in *Granta, Frieze, Sight & Sound, The Guardian, The Washington Post*, and elsewhere. Her short films have screened at the BFI Flare festival and others venues across Europe.

Jo Langdon has a PhD from Deakin University, Australia, where she teaches in writing and literature. Her scholarship, published in journals such as *Critique* (US) and *Current Narratives*, has focused on magical realism, trauma, and elegy. She is also a prize-winning poet and fiction writer.

David McCooey is an award-winning poet, critic, and editor. He is the author of *Artful Histories: Modern Australian Autobiography* (1996/2009), and he has published widely on life-writing and poetry. He is the Deputy General Editor of *The Literature of Australia* (2009), and his most recent book of poems is *Star Struck* (2016). He is Professor in Writing and Literature at Deakin University in Victoria, Australia. His website is davidmccooey.com.

Maria Takolander is a poet, fiction writer, and scholar. Her most recent book of poems is *The End of the World* (Giramondo 2014). A widely published scholar of magical realism, she has also published numerous essays theorising creativity in relation to mental illness, embodied cognition, and sociomaterial practice. She is an Associate Professor in Writing and Literature at Deakin University in Victoria, Australia. Her website is mariatakolander.com.

Gillian Whitlock is a professor in the School of Communication and Arts at the University of Queensland, Australia. Her most recent book is *Postcolonial Life Narratives: Transits of Testimony* (OUP 2015), and she is currently writing a book on the life narratives of asylum seekers in detention, based on the archives of letters and artefacts from Nauru now held at the Fryer Library.

INTRODUCTION

The Limits of Life Writing

As my co-editor Maria Takolander writes elsewhere in this collection, 'Life writing has long been theorised in terms of its limits'. Indeed, one might say that a concern with limits brought the field of life-writing studies into being. The rise of auto/biography studies (the forerunner of life-writing studies) in the 1970s and 80s was in large part a concern with the generic and disciplinary limits of what constituted both auto/biography and 'Literature'. This was despite Paul de Man's warning that attempts to define autobiography in terms of genre 'seem to founder in questions that are both pointless and unanswerable' (919). Philippe Lejeune sought to circumvent such definitional problems by attending to autobiography as a mode of reading, and (famously) understood the relationship between autobiographer and reader as a 'pact' (a formal agreement of *limitations*). Lejeune's legal metaphor and structuralist approach, though, was far from reductive. His conclusion that autobiography is a 'historically variable *contractual effect*' (30) effectively draws attention to the limits of proposing limits.

Looking back at these and other foundational works in life-writing studies, we see limits constantly coming into conceptual play: the limits between literary and factual writing; between narrative as a literary device and narrative as lived experience; and between autobiography and fiction. As titles such as *Fictions of Autobiography*, *Being in the Text* and *Artful Histories* suggest, earlier studies of autobiography habitually dealt with the limits of life writing with recourse to hybridity, if not oxymoron and paradox. And, of course, life writing as a practice, rather than a discipline, has also always been concerned with limits: the limits between self and other, memory and forgetting, past selves and present self, and so on. As Christopher Cowley writes, in *Philosophy and Autobiography*, Rousseau's *Confessions* (1782), conventionally seen as a foundational autobiography, is (among other things) 'a serious exploration of the limits of truthful self-representation' (1–2).

The shift in the last two decades from auto/biography studies to life-writing studies was notably informed by feminist and postcolonial theory, as well as the rise of cultural studies, as seen in the critical attention given to auto/biographical subjects previously silenced, such as women, people of colour, Indigenous peoples, and (more recently) children. Such life-writing theory began from a critique of the human subject (simultaneously universalised and limited as male, European, self-present and autonomous), reconfiguring subjectivity as diverse, provisional and intersubjective. The life writing of such subjects was seen to deconstruct the supposedly secure limits of selfhood and auto/biographical expressions of selfhood.

The move from auto/biography studies to life-writing studies has therefore involved expanding the object of study from putatively literary texts to life narratives as they might be most broadly understood: testimony; autoethnography; digital life writing; and so on. This move has allowed for the consideration of graphic, audio-visual and transmedial forms. These include graphic memoir (or comics more generally), photography, auto/biographical film and video and social media. Indeed, far from being generically 'impossible', as de Man would have it, the generic limits of life narrative have been simply, and constantly, expanding, as seen in the list of life-writing genres in Sidonie Smith's and Julia Watson's influential *Reading Autobiography*. The 52 genres listed in the 2001 edition rose to 60 in the 2010 edition. No doubt there are material reasons for this. As Anna Poletti and Julie Rak write in *Identity Technologies* (a collection of essays that brings together life-writing studies and new-media studies), technological changes lead to subjective-generic ones:

> In the same way that the genres of life writing such as memoir or diary create the terms within which people create identities, Internet affordances can work (sometimes covertly) to create the terms for identification and the rules for social interaction (5).

Generic expansion has been accompanied by concomitant theoretical developments. The various 'turns' (ethical, post-human, affective and so on) that life-writing studies has encountered have characteristically involved attending to extra-literary discourses – often legal and medical – in their consideration of the limits of (the expression and representation of) subjectivities. Projects that have productively engaged with extra-literary discourses include those that address ethics (Thomas G. Couser), human rights and testimony (Kay Schaffer & Sidonie Smith, Gillian Whitlock), and childhood and trauma (Douglas).

If the concept of the limit (or its deconstruction) has been so important to life-writing studies, why, then, draw attention to it now, so late in the game? We have focused on the limits of life writing in part because, while the concept (or trope) is crucial to life-writing studies, its thematisation has often only been implicit and/or part of a wider project, as suggested above. We wish, then, to explicitly draw attention to a central strand of thinking in life-writing studies. Another reason to attend to the limits of life writing in a programmatic way is because it is timely to do so. This is in part, as already noted, because of ongoing material changes in auto/biographical cultures.

The rise of social media, to choose one example, has made auto/biographical representation ubiquitous. Such forms of representation are often the source of considerable cultural anxiety. Social media, mobile networks and smart devices mean that life writing is not only ubiquitous, but also instantaneous and trans-medial. A number of the contributors to this collection attend to life narratives and social media. Kate Douglas's essay on Malala Yousafzai, the Pakistani female-education activist and Nobel laureate, extends Douglass's ground-breaking 2016 study of life narratives and youth culture (co-authored with Anna Poletti), by analysing the importance of Malala's 'collaborative life writing archives', the source of which is an anonymous BBC blog that Malala began writing when she was 11 years old. By considering the (self-)mediation of Malala through various print and electronic auto/biographical narratives, Douglas moves beyond critique to consider how the 'genre-crossing work' of life narratives can be politically and personally productive. She also highlights a long-standing limit of life-writing studies (the adult subject) by focusing on life narratives by a child.

The timeliness of this collection can also be seen in terms of other trans-medial forms and contemporary material events. For instance, the current refugee crisis can be felt in several essays collected here, such as Gillian Whitlock's essay on the graphic journalist Joe Sacco. By focusing on graphic memoir, and a moment in which a lacuna in Sacco's biographical record is momentarily filled, Whitlock's essay offers insights not only into Joe Sacco and his graphic journalism, but also the ways in which life writing's 'limits' can be found in narratives concerning migration and geo-political history. Despite Sacco's description of his work as 'stale journalism', Whitlock's essay is especially timely for Australian readers, given how it situates Sacco's transnational biography and graphic journalism in an Australian context, one that links the historical White Australia policy with the Australian government's current punitive policies and regimes concerning asylum seekers.

Migration and diasporic identity are also central to a growing body of second-generation migrant life-writing. This can be seen in Anne Brewster's essay on *Her Father's Daughter* (2011) by the Australian writer, Alice Pung. Brewster reads Pung's memoir in terms of 'postmemoir' and diasporic writing, arguing that the memoir 'demonstrates how the relatively new body of minority life writing is extending the parameters of the field of life writing by reconfiguring our understanding of transnational memory and historicity within the white nation'.

Brewster's essay, and Pung's memoir, both deal with the trauma of history as a limit of auto/biographical expression. Adam Brown's essay on Holocaust life writing attends to a historical trauma that is often seen as a limit case for not only life-writing practice, but also historical understanding and human experience generally. Exposing another limit in the field, he focuses on so-called 'privileged' Jew and the problem of judgement that arises in reading texts by or about them. For Brown, Holocaust life writing necessitates an 'ethics of reading'.

Jo Langdon similarly attends to historical trauma, in this case the trauma of the Vietnam War. Langdon's case studies – two genre-defying works by the American author (and Vietnam veteran) Tim O'Brien – paradoxically bring together autobiographical writing and the fictional mode of magical realism. O'Brien's assertions about the 'authenticity' of these fantastically inclined works certainly test the limits of life writing. However, Langdon considers the ways in which such oxymoronic writing may offer powerful representations of the 'unrepresentability' of trauma.

Juliet Jacques's essay similarly meditates on the 'blurring of boundaries between autobiography and fiction' – though her focus is on trans life writing. The contemporary public discourse concerning non-binary sexual identities is certainly relevant to any consideration of the limits of life writing. Emergent forms of LGBQTIA life writing highlight discursive limits by deconstructing heteronormative assumptions and mainstream representations of non-heterosexual identities. Here Jacques conceptualises trans life writing as a form of resistance (another concept cognate with the idea of limits), meditating on the aesthetic and political rewards of challenging realist conventions in trans life writing.

For many theorists of life writing, the phenomenology of creative practice itself has represented a limit. Writing as both a critic and a life-writing practitioner, Jacques represents a growing body of scholar/life-writers who breach this limit. Maria Takolander's essay on confessional poetry also illustrates this breach. Takolander, a writer of confessional poetry herself, dismantles critical responses to the American confessional school of poets that valorise pathology as key to poetic creativity and practice. Takolander demonstrates how even a genre as putatively 'authentic' as confessional poetry finds its limits and affordances through its intersubjective and intertextual condition, and through the techne of poetry itself. In addition, she draws attention to a form of writing – poetry – that has traditionally been at the margins of life-writing studies.

Creative practice itself is represented by Oliver Driscoll's 'Reflection', a work of creative nonfiction that brings us back to our starting points: the fundamental concern over the relationship between life and writing; the aestheticisation of experience; and the narrativity (or otherwise) of lived experience. The stylisation of Driscoll's piece also expresses a basic tension between authenticity (a valorisation of embodied experience) and mediation (the simulacrum that any experience is necessarily part of), a tension central to a consideration of the 'limits of life writing'.

Maria Takolander and I wish to thank Maureen Perkins for her patient support, expertise, and enthusiasm for this collection, which comes out of 'The Limits of Life Writing' symposium held at Deakin University, in Geelong, Victoria, in February 2016. The keynote speaker was Gillian Whitlock. I, and my fellow editor Maria Takolander, wish to thank all those who made that event the success that it was. The symposium was auspiced by the Contemporary Histories Research Group (CHRG) at Deakin University, a network of researchers, led by Professor David Lowe, concerned with the history that is still 'with us'; that is, with the unfinished business of the past. The sponsoring of the event by the CHRG was both fortuitous and welcome. Life writing is, after all, key to understanding how the past reverberates in the present, how 'history' is always 'contemporary'. In addition, like history, life writing is a discourse that generates considerable insight, and sometimes anxiety, about its generic and literary status, about the limits between 'life' and 'writing'.

Of course, the concept of a limit remains real, even if that reality only pertains to something as banal as a word limit. There is always more to say, as shown by the impressively inclusive recent anniversary edition of *a/b: Auto/Biography Studies* (on 'The Future of Auto/Biography Studies'). But to reach the limit of this discussion, I will simply reaffirm that life writing and life-writing studies are bound to limits – whether they are to do with national literatures; generic boundaries; types of media; subjectivities; bodies; or the fundamental processes of memory and forgetting. As we hope these essays demonstrate, limits – and the crossing of those limits – are central to the practice and understanding of life writing.

References

Couser, Thomas G. *Vulnerable Subjects: Ethics and Life Writing.* Ithaca: Cornell University Press, 2004.

Cowley, Christopher. "Introduction: What is a Philosophy of Autobiography." *Philosophy and Autobiography.* Ed. Christopher Cowley. Chicago: University of Chicago Press, 2015. 1–21.

De Man, Paul. "Autobiography as De-facement." *MLN* 94.5 (1979): 919–30.

Douglas, Kate. *Contesting Childhood: Autobiography, Trauma, and Memory.* New Brunswick, NJ: Rutgers University Press, 2010.

Eakin, Paul John. *Fictions in Autobiography: Studies in the Art of Self Invention.* New Jersey: Princeton University Press, 1985.

Jay, Paul. *Being in the Text: Self-Presentation from Wordsworth to Roland Barthes.* Ithaca: Cornell University Press, 1984.

Lejeune, Philippe. "The Autobiographical Pact." *On Autobiography.* Ed. Paul John Eakin; trans. Katherine Leary. Minneapolis: University of Minnesota Press, 1989. 3–30.

McCooey, David. *Artful Histories: Modern Australian Autobiography.* Cambridge: Cambridge University Press, 1996.

Poletti, Anna, and Julie Rak. "Introduction: Digital Dialogues." *Identity Technologies: Constructing the Self Online.* Ed. Anna Poletti and Julie Rak. Madison: University of Wisconsin Press, 2014. 1–22.

Schaffer, Kay, and Sidonie Smith. *Human Rights and Narrated Lives: The Ethics of Recognition.* New York: Palgrave Macmillan, 2004.

Smith, Sidonie, and Julia Watson. *Reading Autobiography: A Guide for Interpreting Life Narratives.* Minneapolis: University of Minnesota Press, 2001/2010.

Whitlock, Gillian. *Postcolonial Life Narrative: Testimonial Transactions.* Oxford: Oxford University Press, 2015.

David McCooey

ARTICLES

Joe Sacco's Australian Story

Gillian Whitlock

ABSTRACT
Although Joe Sacco is frequently present in the frame of his comics journalism, as a witness, listener and scribe, he rarely attaches his own autobiographical experience to these representations of self. Recently some more detailed biographical detail about Joe Sacco's own life story has begun to emerge in the frames of his comics, particularly in his work on refugees and asylum seekers. One of the least significant and little known facts about Joe Sacco's life, his childhood as a migrant in Australia, becomes relevant here, extending his enduring commitment to ethical spectatorship, and the visibility of human rights violations, by engaging with this most difficult and intimate work of interrogating citizenship, our own and 'others'.

The limits of autobiography are open to negotiation, and on the move. We see this, for example, in differences between the two editions of the primer for life narrative teaching and research, Julia Watson and Sidonie Smith's *Reading Autobiography: A Guide for Interpreting Life Narratives*. In the decade between the first edition of 2001 and the second in 2010 the field expanded exponentially, they remark, identifying burgeoning sites of life writing in digital and visual media, innovative forms of memoir, and a 'welter' of new theoretical approaches (xi). Two of these millennial developments that recalibrate the limits of autobiography are relevant to the focus of this article: the self-representations of the graphic memoirist and journalist Joe Sacco. The first of these is autographics:

> Autographics: Life narrative fabricated in and through drawing and design using various technologies, modes, signs and materials. A practice of reading the signs, symbols and techniques of visual arts in life narrative. See also autobiography, biography, testimony, comics, self-portrait, avatar. (Whitlock & Poletti i)

The second is the importance of human rights storytelling and issues of trauma, testimony and acts of witnessing in life narrative now. Both of these are instrumental in the craft of Joe Sacco's comics journalism, and here I want to focus on how Sacco draws himself on the page in his journalism as a 'cipher' (Cooke np). Although this cipher is frequently present in the frame of Sacco's comics journalism, as a witness, listener and a scribe, Sacco rarely attaches his own autobiographical experience to his frequently self-deprecating representations of self. However, recently some new biographical detail about Joe Sacco's own life story has begun to emerge in the frames of his comics. In this article I want to pursue what seems to be one of the least significant and little known facts about Joe Sacco's life, his childhood in Australia, and to suggest how this amplifies his graphic art, and deserves more than a passing footnote in Sacco scholarship.

In Figure 1 is Joe Sacco (who we rarely see in 'real' life, as he presents himself autobiographically almost exclusively via the cartoony self suspended here) captured in an

Figure 1. 'Joe Sacco, Cipher and Corgi'.
Source: Stuart Mullenberg, https://www.pdxmonthly.com/articles/2009/11/9/joe-sacco-1209. Permission granted for reproduction.

intimate moment, as the artist looks upon this creation of himself, this 'cipher' that is animated as we read across the frames and the gutters of the comics.

> People tell me that I draw myself in a grotesque way, and I kind of nod my head and agree, but it was very unintentional … The thing is, you can draw yourself the way you look or you can draw yourself the way you feel. I sort of fall into that latter category. I often feel the way I'm drawing myself. There is an accuracy in that drawing (Sacco & Mitchell 54).

Labelling this autographical figure as 'cartoony' emphasises how Sacco moves away from realism in his self-depiction, to create a cipher immersed in affect: 'I often feel the way I'm drawing myself.' In the frames of his comics, this cipher bends and twists, expands and contracts; sometimes a glimpse of a body part appears in the corner of a frame. Here we see just the head and shoulders. Is this figure looking back at Sacco? Is there a mutual recognition? Is there an element of wonder? Scott McCloud suggests in his book *Understanding Comics* that cartoon drawing both amplifies and simplifies: 'When we abstract an image through cartooning we're not so much eliminating details as we are focussing on specific details. By stripping down an image in cartoon drawing an artist can amplify meaning in a way that realistic art can't' (30). McCloud focusses in particular on the force of cartoon drawings of the face. The more cartoonish the face becomes, he argues, the more it becomes an icon that has the capacity to produce recognition and association, to move us and create an empathic attachment. But an empathic attachment to who or to what? And why does this seeing 'I' of the comics have eyes that remain hidden?

In all weathers and situations, Sacco's eyes are masked by the rounded lens of his spectacles. His comics feature graphic depictions of faces that individualise and humanise, with eyes full of emotion. But these are always the eyes of others, those he listens to, never the witnessing 'I' of Sacco himself. Theories about Sacco's spectacles, and what this says about the ethics of spectatorship in his comics, vary. Wendy Kozol speculates that this draws attention this character's role as a voyeur, and to the media's myopic and privileged perspective as outside observers, for Sacco is always looking on professionally, he is a journalist, he is working at his craft (167). Hillary Chute reads this blankness amidst so much visual elaboration in his comics as a sign of the political and aesthetic pressure on the act of seeing and witnessing at scenes of trauma and suffering, and as a surface mark of Sacco's desire to cede the stories he solicits as a journalist to others 'to highlight, superficially at least, a modesty through formlessness in the face of others' experiences' (Chute 337). This suggests Sacco is drawing to create an empathic attachment not on his own account, but for those to whom he bears witness. James Chandler questions whether this blindness represents a kind of Homeric blindness that allows him to see things that others can't see: 'Is it the idea that his eyes reflect rather than see the world? Or are these empty spaces peepholes for us ourselves to see the world afresh?' (Sacco & Mitchell 53). Characteristically Sacco's own comments on this are more direct and self-effacing, and affirms that desire to remain 'nondescript':

> It's deliberate now … But it certainly wasn't in the beginning … But some people have told me that hiding my eyes makes it easier for them to put themselves in my shoes, so I've kind of stuck with it. I'm a nondescript figure; on some level I'm a cipher. The thing is: I don't want to emote too much when I draw myself. The stories are about other people, not me. I'd rather

emphasize their feelings. If I do show mine – let's say I'm shaking (with fear) more than the people I'm with – it's only ever to throw their situation into starker relief.' (Cooke)

Remembering McCloud's comments that the cartoony face is open to recognition, we see here Sacco's desire to create a graphic version of self as cipher, a witness who does not emote too much, that works to create an empathic engagement with the stories about other people.

How does this work?

Figure 2. 'Pilgrimage' from Joe Sacco, 'Palestine'.
Source: Joe Sacco, 'Palestine' 235. Copyright Fantagraphics. Permission granted for reproduction.

In *Palestine*, Sacco draws the intersection of ethics and aesthetics and the visualisation of witness into the frame. *Palestine* was published as a book in 2001, and it gathers a series of nine stories that take place over a two-month period in late 1991 and early 1992, with flashbacks to the expulsion of the Arabs from the West Bank, the Intifada and the Gulf War. Sacco spent much of his time in Palestine in dialogue with Palestinians, narrating their daily struggles and the humiliations of life in the occupied territories, and most particularly, in the refugee camps. These stories were first published serially in comics in 1993.

So, for example, in Chapter 8, 'Pilgrimage', Sacco's fixer, Sameh, has taken him to the home of an elderly Palestinian woman, who has recently lost two sons in violent encounters with Israeli troops. We see Sacco's hands on the edge of the frame (Figure 2), captured in the act of witness; his hands are magnified, copying the testimony he is hearing: 'I get my pen out, and we get right to business' (235). The word 'business' is a reminder: Sacco is on the page as a graphic journalist at work.

This explicit depiction of bearing witness to the dispossessed is drawn on the page: as readers/viewers we access their story from their own mouths – or not quite, because Sacco includes the presence of Sameh, his translator and fixer. Although Sacco himself morphs as he enters the frames of the comics in *Palestine*, the people themselves are always as we see them here: captured in graphic realistic detail. These are historical: they are often named, and are drawn so as to be recognisable as individuals. The accounts we hear are from the Palestinian people themselves, in dialogue with Sacco. In 'Pilgrimage', there are six pages of testimony by the woman who is the centre-point here. She has given this testimony many times we know (it's 'business'), and here she knowingly and deliberately gives it again to the journalist Sacco. She tells the story in the presence of family. Hospitality is evident: tea on the small table – Sacco is awash with tea during this time on the West Bank. When they are finished, she challenges him: 'She wants to know,' Sameh translates, 'how talking to you is going to help her. We don't want money, she says, we want our land, our humanity' (242).

'Pilgrimage' begins with a splash page that illustrates Sacco's journey into this refugee camp. Lacking margins or frames, the entire page is covered with painstaking cross-hatch, shading and stipple. Almost a third of this image is tumultuous sky: huge clouds that buffet the square voice balloons strung across the top of the page to tell the story in Joe's voice:

> There's been some bad news ... Sameh wants to press on regardless ... I'm not insistent ... We could skip the next item on my agenda ... but Sameh is determined to continue ... We've hitched a lift ... a donkey cart ... another authentic refugee camp experience ... good for the comic, maybe a splash page. (217)

And this is what we see. The work of mediation is evident. Here, too, the captions present an ironic commentary – an irony directed to the business of journalism. The comics journo always has an eye for the potential, for the story and the image. Here is an embedded questioning of the ethics of journalism and the search for authentic story.

Yet, at the same time as the splash page voices this questioning of the ethics of journalism and search for authentic story in the experiences of refugees, the picture tells a story in graphic detail: below the turbulent sky we see drawn with painstaking care the donkey cart, the damaged buildings, a perilously poised tower, the rubble that surround the Palestinians, the thick mud that cakes their boots. The keffiyeh and veils that cover the heads of the people blow in the wind, and it is cold and bleak. And there is Sacco himself, carried into the camp on a wagon along with other commodities: trays of vegetables and provisions. In this cross-hatching and these ink-full pages we see an ethics of care, a scrupulous attention to bearing witness to social suffering.

The Palestinian critic Edward Said wrote a wonderful 'Homage to Joe Sacco' that opens *Palestine*, and here he comments in particular on Sacco's graphic rendition of Gaza and its refugee camps. You see all of this 'through Joe's own eyes as he moves and tarries among them,' Said comments, 'attentive, unaggressive, caring, ironic and so his visual testimony becomes himself, so to speak in his own comics, in an act of profoundest solidarity' (v). Typically, in his self-deprecating and ironic fashion, Sacco draws himself reading Said's *Orientalism* and he uses this to question what his seeing 'I' can do as it enters the refugee camps on the west bank – but only briefly, because his cipher has little interest in engaging with theory. Said remarks on seeing '*through* Joe's own eyes' (my emphasis), but as we have seen Sacco's presence problematises looking, seeing, spectacle in every cartoon representation of himself on the page.

In his Editor's Introduction to a fine new collection of essays on Sacco's work, Daniel Worden identifies two traditions in twentieth-century art and literature that are critical in the pedigree of Sacco's comics journalism. The first is the underground comics movement in the USA in the 1960s and 70s, which is critical in the development of autographics as Jared Gardener has argued. These underground confessional comics featured the work of Justin Green, Art Spiegelman, R. Crumb and, as the feminist critic Hilary Chute reminds us in *Graphic Women*, women artists such as Phoebe Gloeckner and Lynda Barry. These comics incessantly return to the artist: their sexuality; obsessions; relationships; and traumatised relations with others. By the late 1980s Sacco was part of this underground comics scene, working for Fantagraphics books, although never in the confessional, highly subjective mode of these early autographics. In fact, Sacco tells us little about his own life history – we know nothing of Sacco's obsessions and passions, and we know very little of his personal life. As in the photograph of Sacco and his drawn self (Figure 1), he remains at one remove from his masked cipher.

The second influence on Sacco's graphic journalism is contemporaneous with the underground comics in the USA: new journalism, and its style of literary reportage that combines the historical and the personal, and licenses that subjective reportage that Sacco practises. In *Journalism* Sacco comments specifically on this art and craft, which he calls an 'adamant art' that draws on these earlier literary and artistic movements of the late twentieth century, and puts them to work in disaster zones:

> In short, the blessing of an inherently interpretive medium like comics is that it hasn't allowed me to lock myself within the confines of traditional journalism. By making it difficult to draw myself out of a scene, it hasn't permitted me to make a virtue of dispassion. For good or ill, the comics medium is adamant, and it has forced me to make choices. In my view, that is part of its message. (*Journalism* xii)

Hillary Chute gets Sacco's autographics right in *Disaster Drawn*, where she identifies this 'adamant art' as a visual work of witness, which uses the word and image form of the comics to create a record of the testimony of the dispossessed that is witnessed on and through the page. 'The medium of the comics,' remarks Chute, 'is always already self-conscious as an interpretive, and never purely mimetic, medium. Yet this self-consciousness, crucially, exists together with the medium's confidence in its ability to traffic in expressing history' (198). In *Journalism*, Sacco talks of this as a 'felt' presence:

> I, for one, embrace the implications of subjective reporting and prefer to highlight them. Since it is difficult (though not impossible) to draw myself out of a story, I usually don't try. The effect, journalistically speaking, is liberating. Since I am a 'character' in my own work, I give myself permission to show my interactions with those I meet. ... In the field, when reporting, a journalist's presence is almost always felt. ... By admitting I am present at the scene, I mean to signal to the reader that journalism is a process with seams and imperfections practiced by a human being. (*Journalism* xiii)

In Sacco's graphics, an ethical reflection on the right to look and the witnessing gaze is embedded in the art. Whereas the spectator gazes passively at violence, the witness undertakes an ethical look that mobilises the viewer's sense of responsibility (Kozol 166). The question of how visual cultures can mobilise emotional and political reactions and open up possibilities for ethical spectatorship plays out on the page of Sacco's graphic journalism. What is ethical vision? A recognition of the limits of what we see and what we know, and the complicity and responsibility of bearing witness to testimony, as Kozol suggests. Sacco inserts his cipher at a critical site/sight: he is there in the position of the intermediary Western figure that, in human rights literature, is the privileged witness that authenticates trauma story and, in the context of journalism, the reporter whose visual witness to human rights abuses is deeply implicated in commercial imperatives for violent spectacles that structure contemporary news reportage or 'Business' (169). Sacco inserts his masked voyeur here.

In the ethical engagement proposed by Sacco's drawing, a slow journalism, or stale journalism as he calls it, there is an extended process of listening and researching, imagining and visualising and finally materialising of other people's experiences – graphically. Pen and paper and hand are instruments of witness, and acts of testimony and witnessing find unique shape in this form (Chute 2) as biography and history are not simply reflected 'but are reinscribed, translated, radically rethought and worked over by the text' (Felman, cited Chute 33). The basic hand-drawn grammar of the comics – frames, gutters, lines and borders – is a unique and graphic textualisation drawn by the hand that bears witness. In the painstaking stipple and cross-hatching of this 'slow journalism' that is always in the aftermath of the breaking story, is an ethics of care.

Journalism presents a manifesto of Sacco's art and craft – or as much of a manifesto as we are ever likely to get from him. It returns to the representation of refugees specifically, and here Joe Sacco breaks cover and writes for the first time autobiographically about 'my own people'. Sacco is often taken as the representative American in *Palestine* and *Safe Area Gorazde*. In 'The Unwanted', first published in the periodical press in two parts in 2010 and then reprinted in the collection *Journalism* in 2012, he reports on the surge of African asylum seekers to Malta, where he was born in 1960. 'I thought,' he tells us, 'there was no better place to report on the issue of African migration to Europe than my own birthplace, Malta … as a Maltese I figured local people would be less reticent with me about their feelings toward the Africans who had landed on the island' (xx). Sacco visited Malta in 2009, just before the Arab Spring, and he maps out what we now know to be the first phase of an exodus of asylum seekers into Europe from Somalia and Eritrea, Nigeria, and West Africa. In 2009, he tells us, 12,500 mostly sub-Saharan Africans washed up on the island's shore (109). Sacco reappears in his familiar cartoon incarnation in 'The Unwanted', though we see a concession to the Mediterranean climate, as the beanie and jacket are replaced with his panama hat and casual shirt. This character is immediately recognisable, but his positioning on the page as a witness is radically different here in Malta.

As Hillary Chute points out, characteristically Sacco has no autobiographical or familial connection to the testimonies he documents. Unlike Art Spiegelman, Alison Bechdel and Marjane Satrapi, he has no identitarian attachment to the communities he enters as a graphic journalist at work. His status as neither Jewish nor Muslim helped him navigate the borderlands in Israel, Palestine and Bosnia. He is present on the page as a professional journo. His work is not about figuring out how the past shapes his own present and nor does he make injury and social suffering the site of his own politicised identity (Chute 205).

Sacco has been criticised for limitations in his approach to historical witnessing in *Palestine*. For example, he doesn't introduce the United States' support for Israel, and in *Footnotes in Gaza* (where we briefly glimpse his Maltese passport) there is only a muted commentary on the role of American citizenship in his witnessing gaze (Kozol 187). He has been reticent to engage with his character's nationality and privileges of citizenship. This changes dramatically in 'The Unwanted', where he introduces a twinned migration journey, in the form of a confession: Sacco's act of 'coming clean' is an admission of the privilege of his own family's emigration, its access to an approved transnational mobility that is very different to the detention of the undocumented who wash up on the shore of Malta (Shay 242).

In 'The Unwanted', where he returns to his homeland in Malta, he introduces his Australian Story in a single panel. What we know of Joe Sacco's biography is limited. He was born in 1960 on the island of Malta. His family emigrated to Melbourne in 1961; his father was a member of the Malta Labour Party, and worried about the influence of the Catholic church on Malta (Cooke), but, he recalls, there were economic reasons for leaving Malta too, and the family relocated to the suburb of St Albans in Melbourne.[1] Sacco's childhood was spent in Australia until 1972, when his family moved to the United States (Worden 4). In Monica Marshall's biography of Joe Sacco, Sacco's Australian story is fleshed out a little: the family lived in a 'small suburb' of Melbourne and his mother was a high-school teacher, his father an engineer. As his biographer, Marshall pays only cursory attention to his childhood. Sacco recalls the family was part of a multicultural community; all his friends were European-born or had European parents: 'In Australia, there were a lot of Europeans and they would all meet up and the commonality was the war. You heard a lot about it. I guess I realised conflict was just a part of life' (Cooke np). As a child Sacco recalls he read British war comics his father bought him in downtown Melbourne; in his family his mother's memories of the bombing of Malta during the war remained vivid. Sacco drew his first comics in the suburbs of Melbourne, aged 6 or 7. His childhood memories of the Vietnam War, and the 1967 Six-Day War in the Middle East when Israel took control of the West Bank, are recollections of discussions with his friends in Australia. The Sacco family left for the USA in 1972, when he was 11, a reminder of their access to the privileges of transnational mobility and settlement.

This is a skeletal biographical sketch of Sacco's childhood. If anything, it seems that Sacco's Melbourne childhood confirmed his identity as European and Maltese, a member of a global diasporic community. Nevertheless, this panel opens an Australian story, and this takes the form of a confession: 'Time for me to come clean,' reads one of the floating captions that taps into tropes of waste and contamination that recur in 'The Unwanted', where the asylum seekers are associated with waste product that contaminate the homogeneity and sovereignty of the island. 'My family immigrated to Australia when I was a baby,' he tells us, ' … and the Australian government, eager to populate its large continent with white-faced Europeans paid most of our passage'. In fact, Maltese had not always been recognised as white-faced Europeans under the terms of the White Australia policy. When the Sacco family came the Maltese were included in those eligible for assisted passage, and the largest Maltese-born community was in Victoria, focussed in the suburbs of Sunshine and Keilor. Sacco's Australian story is, then, part of a longer Maltese story of immigration and resettlement. Who counts as a white-faced European is not fixed in any essential way, and it is subject to politics, and policies of population management. Maltese people were excluded by the 1901 Immigration Restriction Act, and a quota system was introduced for Maltese immigrants in 1920. In 1948, an assisted passage agreement was signed with Malta, the first agreement Australia made with any country other than Britain, and this resulted in an increase in arrivals in the 1950s and 60s.[2]

The Sacco family migrated to Australia aboard the first passenger ship of the Sitmar line, the *MS Fairsea*. After wartime service in the Pacific and a brief post-war period as a troopship, the *Fairsea* was rebuilt for migrant service in 1949 with basic accommodation for 1800 passengers with the intention of transporting displaced people and refugees from Europe to Australia, contracted by the International Refugee Organisation from 1949 to 1951. In 1955 she was chartered by the Australian government to transport assisted migrants from Europe to Australia and New Zealand, and she holds an important place in the memory of many immigrants to Australia in the post-war period.[3] In this frame from 'The Unwanted' (Figure 3), the *Fairsea* looms large, a massive hull dominating the panel in the middle of the page, with Valetta's ramparts to the right of the ship, farewelled by a fringe of hands that are not identical. This is the departure from Malta. The following panel shows African asylum seekers held in detention on Malta, and the viewer's angle of vision is low to accentuate the dominance of the barracks in the background. This is drawn from within the detention centre, and every figure in this frame looks at us, directly. The two figures in the foreground project hopelessness in their posture, their gaze; the refugees are unwelcome, uninvited, and there is no gesture of hospitality here. On the contrary, there is fear, which is drawn graphically in the top panel. Here an elongated frame captures the fear of refugees on the island, and a moral panic that thrives on rumour and gossip. Sacco divides the panel into three with black gutters, and the words are not in floating captions but contained: 'There is a story going around that has gained traction in Malta. I heard it four times.' and 'An African tells a Maltese police-man, "Keep the boats because one day you'll be in them."' (113) This visualises the fearful prospect that the Maltese people themselves will be dispossessed by the refugees, and driven out to sea in small boats. In the notes on migration at the end of 'The Unwanted', Sacco tells us that, though obviously his sympathies are with the migrants,

> I thought it was incumbent on me to treat the fears and apprehensions of the Maltese people seriously. Few peoples, I'm afraid, are up to the challenge of absorbing large and sudden influxes of outsiders, especially those of a different color. My own people are no better than anyone else. (157)

Sacco was not required to engage with his citizenship in his earlier pilgrimages to refugee camps in Palestine and Eastern Europe. To be sure, in some ways 'The Unwanted' draws on his earlier work on ethical spectatorship and refugee testimony. Here, too, we find multiple interviews and visualisations of experiences of suffering and loss, and recognition of the precarious life of refugees. That earlier engagement with the politics of visibility of refugees in *Palestine* is sustained here, in Sacco's active self-reflexive work on the ethics of bearing witness to the dispossessed. However, 'The Unwanted' demands a further and difficult engagement, with the fears and apprehensions of 'my own people'. When Sacco presents his Australian story as a confessional rite, a 'coming clean', he is extending tropes of waste, toxicity and contamination that recur in his interviews with Maltese citizens, as Shay suggests. The asylum seekers are people 'they cannot bear to look at', and Sacco draws out these hostile views of the Africans in frames that occupy the gaze that renders them inhuman, as waste products of globalisation.

Here Sacco's own Maltese citizenship is politicised, and connected to the social suffering of African asylum seekers as well as the paranoia of his fellow citizens. In 'The Unwanted', Sacco is drawn onto the page more intimately than before, and as a citizen he is complicit, a beneficiary of migration schemes that privilege some racial and ethnic identities over others. This single page is a fine example of the compression and

Figure 3. 'The Unwanted'.
Source: Joe Sacco, 'Journalism' (p. 113). Copyright Random House & Holt Inc. Permission granted for reproduction.

economy of autographics: Sacco captures the complicated emotions attached to the boat in this island nation, and this is also an Australian story. Here, Sacco's Australian migration and childhood become meaningful biographical details for his comics journalism, allowing him to draw out the similarities between these island nations, where sovereignty is tied to visions of a homogenous community of the white nation that demands rigorous border protection. Australia and Malta are very different historically, geographically and socially, but Sacco uses his autobiography to create a more expansive and historical account of migration to draw out that carceral archipelago of islands that are connected by the passages of asylum seekers now: not only Malta and Australia, but also the Pacific islands where Australia sustains mandatory offshore detention camps, on Manus and Nauru. They are all drawn into this frame.

Recently there has been extensive media commentary on shame that questions why stories of human rights abuses in the media fail to elicit an ethical and empathic response to refugees and asylum seekers, and why moral disengagement and states of denial persist. There are two narratives in 'The Unwanted', and this moral disengagement is one of them. Euphemisms, denial of human rights abuses, dehumanising rhetoric that justifies inhumanity, and an absence of responsibility are all components of denial at work. Sacco's Australian story extends his enduring commitment to ethical spectatorship, and the visibility of human rights violations, by engaging with this most difficult and intimate work of interrogating citizenship and creating an autographics form that is adequate to the task. 'The Unwanted' is one of the earliest representations of the current exodus of asylum seekers out of Africa and the Middle East to Europe. How journalism can engage with this, and bear witness, is perpetually in question. In 2016, the Pulitzer Prize for Breaking News Photography went to journalists from the *New York Times* and Reuters for their extraordinary photographs and testimonial narratives that followed refugees in their journeys across Europe in the summer of 2015, that gave a face and a name and a history to specific individuals. The *Guardian* newspaper novelist Richard Flanagan and the artist Ben Quilty created a powerful multimedia essay that witnesses the experience of individual refugees and their families. This is what humanitarian narrative does: it tells the single story that can generate empathic engagement from afar. Ethical questions about the right to look and the witness as voyeur are embedded in acts of testimony and witness, and these questions become acute when we see, or when we are prevented from seeing, asylum seekers. New technologies create new possibilities for 'visual activism', Nicholas Mirzoeff's term for the interaction of pixels and action to make change happen (297). But autographics, too, is drawn into visual activism here on the question of how graphic journalism can respond to forced migration. The forcible return and mandatory detention of asylum seekers in Australia, and the question of how it came to be so cruel in response to forced migration, makes this an Australian story.

Joe Sacco's Australian childhood remains an unfinished story in his work. In *The Great War, July 1, 1916*, Sacco creates one accordion-style foldout drawing, a 24-feet-long panorama of the British forces' experience on the first day of the Battle of the Somme. Sacco acknowledges many influences: the French cartoonist Jacques Tardi's *It Was the War of the Trenches*; Matteo Pericoli's wordless *Manhattan Unfurled*; and medieval art and the Bayeux Tapestry. In his 'Author's Note', Sacco points out that he depicts this scene only from the British perspective, for he is familiar with English-language war literature and Anglo-centric histories; it is these that have 'most seeped into my consciousness'.

This 'Author's Note' begins with an acknowledgement of when this Anglocentric consciousness of the war began to influence him:

> The First World War has loomed large in my psyche since my school-boy days in Australia when every April 25 we commemorated the anniversary of the ANZAC landings at Gallipoli. I was cognizant, even then, that a war dubbed The War to End All Wars must have thrown up such horrors that the survivors believed it was the last word on the matter. My fascination with the war in which armies clubbed each other for year after year over small bits of ground has never abated. (1)

Here again, as with his deft reference to the White Australia Policy in 'The Unwanted', Sacco touches on his own childhood experience of Australian cultural heritage to make a piercing observation about Australian citizenship. The Anzac legend is the focus of a narrative that draws on grief and mourning to create a dominant and Anglocentric account of Australian nationhood. Sacco's boyhood is not a dim memory at all; on the contrary, it emerges in his sharp sense of its legacies in his own psyche, and his perceptions of Australian citizenship and cultural heritage more generally.

Joe Sacco's visual activism is an intricate art. In 'On Satire', his autographic response to the *Charlie Hebdo* killings in Paris in January 2015 that was published in *The Guardian* two days after the attacks, he places himself as a satirist at the front line of debates about visual activism, the right to draw and what can be drawn. To return to that enchanting photograph of Sacco and his icon (Figure 1), we see Sacco holds his drawing up just beneath several conventionally framed etchings on the wall, accentuating just how flimsy and ephemeral that page he creates is. Together, both artist and cipher cast a shadow on the wall. We don't look at this unobserved, because we too are being watched. The eye of the corgi peeps at us from the corner of the frame – always a site of action and witness in Sacco's graphics. This is whimsical – Sacco has drawn a graphic 'Portrait' of his corgi: 'I shudder to think how my future biographers will judge you dog' (*Portrait*). Future biographers will pay more attention to Joe Sacco's Australian Story, and cursory descriptions of a suburb near the beach and an Australian childhood as a blank slate or dim memory will no longer seem good enough. This photograph, like all Sacco's autographics, is a canny self-reflexive art that is powerfully self-contained. All the more reason to look more closely at the presence on the periphery of the frame.

Notes

1. Joe Sacco, personal communication, 8 March 2017.
2. Details of Maltese migration to Victoria are drawn from the Museum of Victoria website: http://museumvictoria.com.au/origins/history.aspx?pid=39
3. Post World War II Migrant Ships: *Fairsea*. https://museumvictoria.com.au/discoverycentre/infosheets/fairsea/

Acknowledgements

Thanks to Joe Sacco and to Grace Dietshe for assistance with this article.

Disclosure statement

No potential conflict of interest was reported by the author.

References

Chute, Hillary L. *Graphic Women. Life Narrative and Contemporary Comics*. New York: Columbia University Press, 2010.

Chute, Hillary L. *Disaster Drawn: Visual Witness, Comics and Documentary Form*. Cambridge, MA: Harvard University Press, 2016.

Cooke, Rachel. "Eyeless in Gaza." *The Guardian* 22 Nov. 2009. https://www.theguardian.com/books/2009/nov/22/joe-sacco-interview-rachel-cooke Accessed 1/11/16.

Flanagan, Richard. "Notes on the Syrian exodus: 'Epic in scale, inconceivable until you witness it'." *The Guardian* 5 March 2016 https://www.theguardian.com/world/2016/mar/05/great-syrian-refugee-crisis-exodus-epic-inconceivable-witness-lebos-islamic-state Accessed 1/11/16

Gardener, Jared. "Autobiography's Biography, 1972–2007." *Biography* 31.1 (2008): 1–26.

Kozol, Wendy. "Complicities of Witnessing in Joe Sacco's Palestine." *Theoretical Perspectives on Human Rights and Literature*. Eds Elizabeth Swanson Goldberg and Alexandra Schultheis Moore. New York: Routledge, 2012. 165–79.

Marshall, Monica. *Joe Sacco*. New York: Rosen, 2005.

McCloud, Scott. *Understanding Comics: The Invisible Art*. New York: HarperPerennial, 1994. http://theconversation.com/robert-manne-how-we-came-to-be-so-cruel-to-asylum-seekers-67542 Accessed 1/11/16.

Mirzoeff, Nicholas. *How to See the World*. New York: Pelican, 2015.

Sacco, Joe. *Palestine*. Seattle: Fantagraphics Books, 2005.

Sacco, Joe. *Journalism*. London: Jonathan Cape, 2012.

Sacco, Joe. "Portrait of the cartoonist as a Dog Owner." *The New York Times* 13 Aug. 2010. Accessed 1/11/16.

Sacco, Joe. *The Great War, July 1, 1916: the first day of the Battle of the Somme: an illustrated panorama*. New York: W.W. Norton & Company, 2013.

Sacco, Joe. "On Satire – a response to the Charlie Hebdo attacks." *The Guardian* 10 Jan. 2015.

Sacco, Joe and W.J.T. Mitchell. "Public Conversation: Joe Sacco and W.J.T. Mitchell. May 19, 2012. Introduced by Jim Chandler." *Critical Inquiry* 40.3 (2014): 53–70.

Shay, Maureen. "'What Washes Up onto the Shore': Contamination and Containment in 'The Unwanted'." *The Comics of Joe Sacco: Journalism in a Visual World*. Ed Daniel Worden. Jackson: University Press of Mississippi, 2015. 239–55.

Smith, Sidonie and Julia Watson. *Reading Autobiography: A Guide for Interpreting Life Narratives*. Minneapolis: University of Minnesota Press, 2010.

Whitlock, Gillian and Anna Poletti. "Self-Regarding Art." *Biography* 31.1 (2008): v–xiii.

Worden, Daniel, ed. *The Comics of Joe Sacco: Journalism in a Visual World*. Jackson: University Press of Mississippi, 2015.

Malala Yousafzai, Life Narrative and the Collaborative Archive

Kate Douglas

ABSTRACT
This article looks at the numerous life narrative texts authored by and written about the Pakistani youth activist Malala Yousafzai. I consider the ways in which Malala's archive of collaborative life narrative texts, which represent a progressive narrative that explicitly addresses critiques made about Malala as it proceeds, makes visible moments of resistance. Malala has crafted a speaking position that utilises the ideologies and cultural constructions of childhood and youth – particularly as citizens and representatives of a nation's future – to become a voice of educational reform.

On the 9 October 2012, the young Pakistani blogger and educational activist, Malala Yousafzai, was shot in the head at close range by the Taliban. After months of treatment and rehabilitation, she made a miraculous recovery and continued her activism. She won the Nobel Peace Prize in 2014. Malala is perhaps best known as a political activist, but her activism stems from her life writing. Her life writing began with the pseudonymous blog that she wrote for the BBC in 2009; she has since written two print memoirs: her co-authored (with Christine Lamb) memoir *I Am Malala* (2013), which was followed by the publication of a 'young readers' edition' co-authored by Patricia McCormick (2016). Malala's life writing has been punctuated by biographical texts about her life. In 2009, Adam B. Ellick, a journalist for the *New York Times*, and Irfan Ashraf made the short documentary *Class Dismissed*, which profiled the experiences of Malala and her activist father Ziauddin at the coalface of the violent opposition to educational inequality in the Swat Valley in Pakistan. The documentary film *He Named Me Malala* directed by Davis Guggenheim was released in 2015.[1] Collectively these texts represent a collaborative archive: a cumulative series of life narrative texts authored by Malala and others which have become authoritative narratives in circulating Malala's life story.

Though it is not unusual for one person to have written multiple life narrative texts (or have had multiple life narrative texts produced about them, or in collaboration), Malala's age perhaps makes the proliferation of life narrative texts more notable, and perhaps more open to critique. Phyllis Mentzell Ryder summarises the critiques that have been made of Malala and, more particularly, of Western responses to Malala, given the willingness to publish her narratives, and to promote her public image: 'Critics, while impressed with Malala and her courage, question why she is a darling of Western media'. (175) Mentzell Ryder quotes commentators who describe Malala as 'a tool for political propaganda' who is being 'used and misused' (175). However, she offers a persuasive argument against these critical positions: 'I am frustrated that the critique ends there ... What distresses me ... is how the analysis focuses only on how Malala has been re-written by the West ... the critique positions "appropriated activists" as helpless victims.' (176) Such critical positions take away Malala's agency and potential for resistance (Mentzell Ryder 176). She argues:

> Instead of noting only *that* appropriation happens ... can critics do more than describe this appropriation? Is it possible to make visible moments of resistance, moments of potential agency? More specifically, when we examine Malala's actions and rhetoric, can we identify any counter-narratives, places where she exceeds the stories told about her? (176)

This article takes up this challenge: to consider the ways in which Malala's archive of collaborative life narrative texts, which represent a progressive narrative that explicitly addresses critiques made about Malala as it proceeds, makes visible moments of resistance.

Shenila Khoja-Moolji usefully refers to the plethora of writings by and about Malala as an 'archive' of Malala (541) – an assemblage of knowledge about her that we might read to better understand her story and the broader contexts and implications of her life. I want to take this idea further through engaging life writing theory to read parts of the Malala archive. Life narrative theorists commonly accept mediation, remediation and collaboration, though challenging and potentially problematic, as an inevitable process in the circulation of life stories (see Douglas and Poletti; Jensen and Jolly; Kurz; Schaffer and Smith; Whitlock – all of whom differently explore the varied processes of mediation and their effects on the production and circulation of life narrative).[2] Jensen and Jolly remind us:

> All life storytelling is mediated and conditioned by the vagaries of memory, particularly when trauma is involved. In the extremely political situation of inflicted suffering and potential reparation, the pressures to shape the story are often as stark as the expectation that they will be only fully truthful. (11)

The challenges that mediation and collaboration bring to life narrative allow us to think through the dynamic nature of life narrative and consider what sorts of stories life narrative genres may enable and limit – often in the same instance. Collaborative archives become meaningful when they make transparent some of the particular ways that life narratives are assembled, circulated, contextualised and interact with other texts. For instance, the multiple life narrative texts of Malala layer new knowledge and speak back to previous texts – particularly critiques of Malala and her father Ziauddin. These life narrative texts anticipate readerships and audiences and speak to their desire for new knowledge and perspectives on and from Malala. In doing so, these multiple, mediated, collaborative life narratives reveal significant knowledge about the limits and affordances of contemporary life narrative genres.

Diary of a Pakistani School Girl (2009)

Malala began speaking publically about her right to an education in 2008 (at age 11) at the Press Club in Peshawar. She was supported by her education activist father, Ziauddin Yousafzai (Westhead). In 2009, Malala started writing a pseudonymous blog for BBC Urdu (the blog was published in Urdu and English) (van Gilder Cooke). The Head of BBC Urdu, Aamer Ahmed Khan, wanted a female teacher or student to blog about her life in the region. Amid the plethora of available sources of knowledge on this issue which include military propaganda, journalism (from different sides), human rights discourse and interdisciplinary 'expert' scholarship, there was a perception that there might be a need for another accessible and authentic angle on the events in Pakistan. Why an anonymous schoolgirl blogger? (The blog is pseudonymous rather than anonymous because Malala adopts the pen name Gul Makai. But it is often referred to as an 'anonymous blog').[3]

The desire to have a young girl narrate the events from Swat signals the importance of gender and youth as important aims for the project and themes for the blog. As I have discussed elsewhere, youthful voices are so often missing when it comes to recording political and historical events, and young women's voices even more significantly absent (Douglas and Poletti). A youth-authored blog potentially fills a gap in knowledge and perspective: prioritising particular subjects such as young women's right to an education in Pakistan, and functions as a call-to-action from someone 'at the coal face'. Indeed, BBC Urdu (and the BBC beyond this) have much to gain from the inclusion of a youthful voice like Malala's: young writers are, conventionally, highly sympathetic and believable. The success of such a blog means an increased readership (and thus economic gain), and an increased cultural capital gained from being associated with the writer and text.

The purpose of the blog was to advocate for young women's right to an education.[4] So,

> Malala chose a pseudonym – Gul Makai, the name of a heroine from a Pashtun folk tale – and began dictating her diary to Kakar weekly over the phone. She described going on trips to buy bangles, living in a place as beautiful as the Swat Valley and the disappointment of being banned from school by the Taliban. (van Gilder Cooke)

The narrative fulfilled certain (anticipated) reader desires – for 'authentic' life stories, and for counter-narratives.[5] So, at this time (2009), two things were happening concurrently: the anonymous blog was achieving a wide readership; and Malala was becoming well known for her advocacy work. Following the release of Adam B. Ellick's and Irfan Ashraf's documentary *Class Dismissed* (2009), Malala became known and began giving interviews and emerging as a public figure on the issue of the right of young women to an education.[6]

Why was there so much engagement with Malala's blog? How did this blog (and how does it) exist as an archive of girlhood experience and now a part of a larger Malala archive? We know that the blog posts were edited and translated and thus mediated for an anticipated readership (perhaps initially Pakistani ex-patriots, and later BBC readers worldwide). The blog underwent various mediations between Malala writing the blog (and perhaps her father reading it and helping in this process) and being received by the BBC contact, being edited, having certain extracts chosen for BBC Urdu, and then extracts chosen for the BBC news site. We cannot know precisely the extent of these mediations, but it is worth acknowledging and speculating on the potential effects of such mediations on our interpretation of the blog.

The BBC news story, titled 'Diary of a Pakistani School Girl' (2009) is annotated with photographs from Swat, the town where Malala lives, and the site offers this introduction:

> Private schools in Pakistan's troubled north-western Swat district have been ordered to close in a Taliban edict banning girls' education. Militants seeking to impose their austere interpretation of Sharia law have destroyed about 150 schools in the past year. Five more were blown up despite a government pledge to safeguard education, it was reported on

> Monday. Here a seventh grade schoolgirl from Swat chronicles how the ban has affected her and her classmates. The diary first appeared on BBC Urdu online.

Contextualisation is important in making the blog accessible to a wide readership, and this annotation invites the reader to consider the blog as one about educational inequalities, but more particularly the Taliban's oppressive impact here and the government's inability to protect young women's right to an education. Thus, the blog immediately invites a benevolent relationship between a knowing, empathetic reader and an innocent victimised young girl.

The blog itself offers 10 short entries, each with a date and title for example: 'Thursday January 15: Night Filled with Artillery Fire.' The two strongest features of Malala's entries are (1) the juxtaposition of her desire for education and descriptions of life in war-torn Swat, and (2) a consciousness of the public nature of her writing and its potential to reach an audience. For example, she writes:

> The night was filled with the noise of artillery fire and I woke up three times. But since there was no school I got up later at 10 am. Afterwards, my friend came over and we discussed our homework.
>
> Today is 15 January, the last day before the Taleban's edict comes into effect, and my friend was discussing homework as if nothing out of the ordinary had happened.
>
> Today, I also read the diary written for the BBC (in Urdu) and published in the newspaper … My father said that some days ago someone brought the printout of this diary saying how wonderful it was. My father said that he smiled but could not even say that it was written by his daughter.

The relationship between everyday life, and everyday girlhood amid terror is important in establishing the particularities of her situation and the potential of the blog to raise awareness of her experience. In other entries, she speaks of not knowing whether the school will reopen for the following school term (grappling with a lack of information). She imagines herself as representing the girls she goes to school with and, in a personal and emotive tone, describes the feelings of longing and loss around her schooling:

> This time round, the girls were not too excited about vacations because they knew if the Taleban implemented their edict they would not be able to come to school again. Some girls were optimistic that the schools would reopen in February but others said that their parents had decided to shift from Swat and go to other cities for the sake of their education.
>
> Since today was the last day of our school, we decided to play in the playground a bit longer. I am of the view that the school will one day reopen but while leaving I looked at the building as if I would not come here again.

In other posts, we are reminded of the young girl's vulnerability to communal trauma as she relates stories she overhears from her parents and from children at school (of death and destruction) as well as her hearing the bombing campaigns. For example, in her post of 3 January titled 'I am afraid', she writes:

> I had a terrible dream yesterday with military helicopters and the Taleban …
>
> Only 11 students attended the class out of 27. The number decreased because of Taleban's edict … On my way from school to home I heard a man saying 'I will kill you'. I hastened my pace and after a while I looked back if the man was still coming behind me. But to my

> utter relief he was talking on his mobile and must have been threatening someone else over the phone.

This post reads like a short story – a thriller with a dramatic climax – but the reader is well aware that this is not fiction, and the emotional and physical vulnerability expressed by Malala is immediate to her. Again, everyday practices (sharing family meals and going to school) are discussed alongside the larger violent conflicts, and it is likely that writing about these subjects and setting them up in contrast was the mandate given to Malala.

The blog consistently criticises the government's and army's inability to protect the schools from destruction and for not prioritising education. For example, on 18 January she writes:

> My father told us that the government would protect our schools. The prime minister has also raised this issue. I was quite happy initially, but now I know but this will not solve our problem. Here in Swat we hear everyday that so many soldiers were killed and so many were kidnapped at such and such place. But the police are nowhere to be seen.

And on 19 January, she writes:

> But the army is not doing anything about it. They are sitting in their bunkers on top of the hills. They slaughter goats and eat with pleasure.

A child's ability to criticise the institutions that affect their everyday lives is often very limited. The internet provides a space to engage directly (as a citizen journalist) and become a more visible stakeholder and activist. Malala's commentary here reveals and engages directly with the power structures that are affecting her. She uses life writing and the form of the blog as a means to write directly of the ways that adults are letting children down, and in doing so, authorises herself as authentic observer of events.

Malala's blog, as published by the BBC, contained 10 entries – about a paragraph each. It is not clear whether this is the sum-total of entries Malala wrote, but her success as a writer and speaker has gone way beyond this. Malala has become a public intellectual and human rights worker who has advocated on behalf of many. As a survivor of violence, she has become an iconic figure of bravery and resilience. Inevitably, though, Malala's rise to prominence and her symbolism to the West as a rebel from the East have been problematised by some critics. Huma Yusuf summarises the three main complaints stemming from Malala's rise to prominence: 'Her fame highlights Pakistan's most negative aspect (rampant militancy); her education campaign echoes Western agendas; and the West's admiration of her is hypocritical because it overlooks the plight of other innocent victims, like the casualties of U.S. drone strikes.' There is recognition here that a blog (or any piece of public life writing that resonates widely) can never be just that. The pen is mighty and Malala's success proved advantageous to powerbrokers with agendas who might recruit her to their cause. For instance, Assed Baig notes in his *Huffington Post* piece 'Malala Yousafzai and the White Saviour Complex':

> Straight away the Western media took up the issue. Western politicians spoke out and soon she found herself in the UK. The way in which the West reacted did make me question the reasons and motives behind why Malala's case was taken up and not so many others … there is a deeper more historic narrative that is taking place here. (Baig)[7]

This is undoubtedly complex terrain. As Gillian Whitlock has argued, life narrative from locations of Western military conflict can easily become 'soft weapons' in the campaign to maintain public support for military intervention and the financial support for foreign governments who are deemed allies of the War on Terror. Clearly, the success of Malala's blog and her public identity can be largely credited to the Western (politically charged) media's championing of her cause. It seems unjust to deny Malala and her father significant agency in this process, as Baig implies: after all, Malala and Ziauddin Yousafzai had been working tirelessly for some time as educational activists in Pakistan. Margaretta Jolly and Meg Jensen remind us that 'testimony requires more than a legal response. Indeed, the life story as a human rights story cries out for emotional recognition and mourning, the compensation of public remembrance and education.' (5). The example of Malala raises fundamental questions for life narrative researchers regarding the transnational circulation of life writing, mediation and the creation of collaborative archives, and how these narratives are utilised by activists to communicate beyond their local communities. Larger questions about universal children's and girls' rights emerge at this juncture. These larger questions and contexts go some way to explaining the conflicting reactions to Malala. For critics like Baig, the veneration of Malala is reductive, he writes:

> The truth is that there are hundreds and thousands of other Malalas ... Many are victims of the West, but we conveniently forget about those as Western journalists and politicians fall over themselves to appease their white-middle class guilt also known as the white man's burden. ... Malala is the good native, she does not criticise the West, she does not talk about the drone strikes, she is the perfect candidate for the white man to relieve his burden and save the native.

Such commentary is persuasive and reminds us of the problematic histories facing many life narrators whose narratives have (not necessarily by their design) become master narratives and perceived to be representative histories, which in turn dilutes other histories and prevents other narratives from being received. Of course, Baig's criticisms here are not centred on the blog of Malala Yousafzai but on the larger reception and symbolism of 'Malala'. This should not necessarily diminish the important life narrative work achieved by Malala, or the textual and contextual reading of the blog that I write about here. Beyond the plethora of political issues and controversies is a young woman who has engaged herself within complex debates affecting her future. We cannot know (conclusively) the extent to which she has exercised agency and autonomy in writing her blog, but such critiques can be read through the circulation of Malala's two memoirs that were produced in the years that followed.

I Am Malala (2013 and 2016)

As much as it might seem inevitable that Malala's memoir is a co-authored one (with Christina Lamb), it also seems quite fitting: this is a book that promises to be a collective, collaborative social justice project from its opening pages. The book is dedicated 'To all the girls who have faced injustice and been silenced. Together we will be heard.' This collective project is emphasised throughout the memoir as Malala never writes about herself in isolation: her experiences are contextualised within traditional folklore, the cultural and political histories of Pakistan, international politics, family life and friendships. The journalist Christina Lamb is Malala's co-author and her name features on the cover of the memoir underneath Malala's, in smaller writing. This is a prominent acknowledgment: many so-called celebrity memoirs do not overtly acknowledge co-authorship. Lamb is not a ghostwriter or even an as-told-to co-writer; Lamb is positioned as a collaborator. Lamb's presence (and indeed kudos) in the memoir is clearly important, even tactical: she is an award-winning journalist with significant credentials as an international writer on current affairs. She is the foreign correspondent for *The Sunday Times* in the UK and worked for many years in Pakistan writing about its politics. She has written four books. (ChristinaLamb.net). Lamb, as a woman of Western origin who has written intelligently and supportively about Pakistani politics, is the ideal conduit for this story and, in particular, for bringing the story to a global readership. Readers may imagine her as fair-minded and objective, while also empathetic towards Malala as a young, politically engaged writer. Thus, the relationship would appear mutually beneficial: Malala is a significant subject and without this subject, Lamb could not put her journalistic writing skills to work. And Malala is able to draw on Lamb's wealth of professional writing experience to assemble a book that is accessibly written and well-structured according to the templates and expectations of contemporary, global publishing industries.

It is perhaps tempting, as Fatima Bhutto in her review for *The Guardian* wonders, to speculate upon how present Malala's voice is in the text. To what extent has Christina Lamb intervened within the text: as writer, facilitator and editor? We know that Malala's story has been consistently critiqued in this way, as mentioned earlier in the discussion of Malala's blog. In the memoir, Malala anticipates such critiques and addresses them overtly:

> It seemed like everyone knew I had written the BBC diary. Some thought my father had done it for me, but Madam Maryam, our principal, told them, 'No. Malala is not just a good speaker but also a good writer.' (51)

Just as Malala asserts her own authorship in the text, *I Am Malala* contains many overt acknowledgements and transparent examples of its being a collaborative text – informed by the perspectives and knowledge of others. For instance, the prologue (and indeed much of the memoir) contains contextual information about Pakistani history and culture (researched by Malala and others). The first few chapters recall Malala's birth and early childhood. The reader feels the presence of Malala's father and mother in these stories. Malala writes of the strong love within her family. We learn that Malala was named after an Afghani warrior – her feminist fate sealed at birth. The text is also careful to position Malala as both remarkable and ordinary. For example,

> I wasn't scared, but I had started making sure the gate was locked at night and asking God what happens when you die. I told my best friend Moniba everything. We'd lived on the same street when we were little and been friends since primary school and we shared everything. Justin Bieber songs and Twilight movies, the best face-lightening creams. (7)

Such statements of Malala's everyday girlhood are important because they anchor Malala in Western childhood and girlhood experience, which in turn emphasises the remarkable nature of her experience and sets her apart. Such contrasts are central to what is expected of auto/biographical texts in Western cultural domains. The production and shape of *I Am*

Malala (perhaps unsurprisingly) reflects a deep consciousness of the genre expectations: it is a text that emerges from cultural systems that (paradoxically) encourage the text to conform, while allowing it to expand the genres of life narrative by making transparent some of the particular processes of influence and mediation affecting the building of a memoir.

The prologue and the latter part of the memoir are concerned with Malala's shooting. The reader is aware that Malala cannot remember much of the events following her shooting and she acknowledges this:

> My friends say he fired three shots, one after another. The first went through my left eye socket and under my left shoulder. I slumped forward onto Moniba, blood coming from my left ear, so the other two bullets hit the girls next to me. One bullet went into Shazia's left hand. The third went through her left shoulder and into the upper right arm of Kainat Riaz. (9)

Readers of memoir understand the ways that memoirs are necessarily constructed from a range of available knowledge and discourse: for instance, traditional knowledge (a chapter epigraph contains a Pashto couplet), the narratives of others, historical knowledge, medical records, and so forth. We can read Malala's description here as a deliberate and sustained means for exposing the limits of life narrative. It is not just her story: her friends were shot, too; many were affected. But also, many may benefit from the telling of this story.

Balance is important here because Malala is careful to emphasise her own agency as a writer and politics as an activist, while revealing the processes of mediation in the text. For instance, part of Malala's project of resistance is resistance to the violence of the Taliban, but just as significant is her critique of narrow Western perceptions of Pakistan and its cultures. She is also critical of Western interventions into Pakistani conflict: 'Malala admits that the Taliban are bad guys. But as much as Malala decries their brutality, she resists attempts to use their story to justify drones or wars. From her perspective, Western military intervention is the exact wrong response' (Mentzell Ryder 181). *I Am Malala* describes in detail Pakistan's beautiful and rich landscapes, cultures and diverse history. She grew from her father's engagement with Pakistani/Afghani folklore, educational advocacy and feminism. These ideologies are set up as complementary. For instance, Malala's reading of the Quran argues for her right to an education: 'The Quran says we should seek knowledge, study hard and learn the mysteries of our world' (154). Later she writes specifically about her religious beliefs and practices, and the strong relationship and synergies between these beliefs and her activism:

> I love my God, I thank my Allah, I talk to him all day. He is the greatest. By giving me this height to reach people, he has also given me great responsibilities. Peace in every home, every street, every village, every country – this is my dream. Education for every boy and every girl in the world. (313)

She learns about Anne Frank and becomes aware that in writing the BBC blog she is part of a literary tradition of young women writing about war (155). She understands her potential contribution to global, cultural history. Khoja-Moolji describes Malala's position here as playing 'at the border'. She explains: 'This liminality of Malala can function as counter-gaze that can effectively displace social control' (Khoja-Moolji 552). Thus, Malala's agency comes precisely from her ability to work within and between different cultural ideologies, histories and languages. Her participation in, and contribution to, these collaborative archives affords her significant power to speak.

The 'Young Readers' Edition' of *I Am Malala* followed in 2016. It is a rewritten version with a new co-author (Patricia McCormick): shorter in length, and with less focus on background and contextual information, and more emphasis on Malala's story (Brien). This edition anticipates a young reader (early teenager and upwards). There is little information about the text's reception, but one educational blogger suggests that the young readers' edition seems more accessible, and Malala's voice seems more apparent in the text. She writes:

> I think the biggest difference between the two books is that the latter sounds more like the Malala we hear in her speeches. The young readers' edition focuses more on Malala's story itself and uses more direct language as well. Not only was it more enjoyable to read, it seemed more like she had actually written it herself .(Kasey Butcher *Ph.D.s and Pigtails*)

Butcher goes on to suggest that there is less emphasis on Malala's father in the young readers' edition and more information about the role that Malala's mother played in educating Malala. Butcher notes how the co-author, Patricia McCormick, takes more of a supporting role in this text than Christina Lamb does in the other edition. (For instance, both Malala and Christina Lamb each have an acknowledgements section in the earlier version; in comparison, the young readers' edition includes only Malala's acknowledgements.) Butcher interprets this as Malala having a more active and central authorial role in the production of the second text. We cannot know definitively whether this is true; however, as Butcher argues, it is certainly constructed as such and this construction is important. Again we can see a layering of knowledge about Malala and her story that anticipates critiques that have gone before: about previous texts and Malala's agency within their production. By focusing more on Malala's voice and centralising her authorship, the young readers' edition adds another element to Malala's life narrative archive, again revealing the extent to which reception concepts have come to layer and shape the archive of Malala.

The focus on Malala's personal narrative and the de-emphasising of contextual information in the young readers' edition is perhaps not so surprising, as it reflects assumptions about what young readers might prefer to, or be more capable, of reading. The *Kirkus Reviews* critique of the book finds this change problematic, takes the opposite view to Butcher's review:

> The story is straightforward, related chronologically … Yousafzai's voice is appealingly youthful, though it often tells rather than shows and frequently goes over the top: In her school, she writes, 'we flew on wings of knowledge.' Still, young Western readers will come to understand the gulf that separates them from Yousafzai through carefully chosen anecdotes, helping them see what drives her to such lyrical extremes. Unfortunately, much is lost in the translation from the adult book, presumably sacrificed for brevity and directness; most lamentable is social and political context. Although readers will come away with a good understanding of Yousafzai's immediate experiences, the geopolitical forces that shape her culture go largely unmentioned except in a closing timeline that amounts to a dizzying list of regime changes.

This argument is persuasive: it is problematic to decontextualise life stories from social, historical and political contexts; such representations could easily be interpreted as a 'dumbing down' of knowledge (which is, in turn, potentially insulting to young readers). However, it is equally persuasive to imagine that this new, different text allows Malala's life narrative to be reinvented again for a different readership. This not only widens the readership potential, but also (again, like the other texts) allows the new memoir to speak back to the previous ones in productive ways.

He Named Me Malala (2015)

In 2015, another life narrative text became part of the Malala archive: the documentary film *He Named Me Malala*. This text, like the others before it, built on both the public knowledge of Malala's experience, and actively responded to critiques of Malala as a public, political force. The title of the documentary immediately reveals the significance and indeed centrality of Ziauddin Yousafzai in both Malala's life, but also in the construction of her story and public profile. 'He' is Ziauddin, and the documentary explains how, in naming Malala after the Afghani folk hero Malalai of Maiwand, who was killed for speaking out against authority (Malalai features in animations within the documentary), her father seems to have predestined her fate.

It is likely that most who view this documentary would already know much about Malala (and indeed this prior knowledge is likely to have led the viewer to the documentary). The interest in her story warrants the circulation of multiple life narratives. Different mediums have the potential to reach wider, even different audiences or readers. Different mediums also have the capacity to shape the narrative differently and extend it in significant ways. And this seems important to *He Named Me Malala*.

This is a documentary that speaks back, explicitly and implicitly, to the previous life narrative texts about Malala. It is generally accepted that documentary is a highly mediated, constructed, creative form – directed, framed and edited for effect. Documentaries usually have a story to tell and use techniques to engage and persuade the viewer to respond in certain ways. *He Named Me Malala* features various techniques designed to add authority to the narrative and to bring us closer to Malala and her experiences: for instance, photographs of the bloodstained bus in which Malala and her friends were shot; documents from life including Malala's brain scan which shows the effects of being shot in the head at close range; archival footage of a younger Malala; video footage of Malala in hospital, including very intimate footage of her post-shooting rehabilitation; interviews with Malala, her parents and siblings in their home in Birmingham; footage of Malala at school; clips of Malala engaged in public speaking, visiting schools in Kenya and refugee camps in the Middle East; and giving interviews on television programmes. *He Named Me Malala* is non-linear and juxtaposes representations of Malala's past traumas with her present triumphs to emphasis her recovery and ongoing successes. This structure is significant because it asks us to re-read the trauma story – as one of forgiveness and survival.

The documentary features a voice-over narration by Malala and the persistent centrality of her voice is important in creating the impression that it is Malala who is telling the story – her story – rather than having it told for her. In using this technique, the documentary speaks back, directly, to claims that Malala lacks agency and/or is being used by her father and Western media to promote certain political agendas (that demonise Pakistan and Islam and centralise the white saviour in Malala's story). It is significant that Malala's voice dominates the documentary and her father speaks only sporadically in interview form (compared with the earlier *New York Times* documentary in which Ziauddin is more central). For instance, Malala is shown teaching her (hapless, technologically challenged) father to use Twitter, and such inclusions are significant in emphasising Malala's agency as life narrator. The documentary includes interviews with critics of Malala who suggest that Malala is a 'character' written by her father. The critics question why Malala's story has so much traction when others are doing more advocacy work in Pakistan? The documentary addresses such critiques indirectly rather than directly; for instance, by following such criticisms with Malala explaining how she became a blogger for the BBC and how her blogs came to be published. There is much emphasis on Malala's voice – on Malala as speaking agent. As Malala says in the documentary, 'There is a moment where you have to choose: to be silent or to stand up.'

The visual narrative allows for a further building of Malala's public persona: as girl advocate. For instance, she speaks directly to camera to explain the teachings of Islam and contrast these to the (non-peaceful) actions of the Taliban. She explains the events of the shooting – drawing a picture on a piece of paper mapping the position she was in on the bus. The narrative is interspersed with her playing a game of cards with the two other girls who were shot at the same time: Shazia Ramzan and Kainat Riaz. Their presence and voices are important as they counter suggestions that Malala's narrative has filled all available spaces for talking about this event and wider event impacting upon these young women, the community, region and Pakistan.

Malala's life in Britain is shown to be difficult: though highly intelligent she struggles keeping up with school work (because her advocacy work keeps her so busy); she reminds us that despite living in Britain, her life will be different to other girls her age (for a multitude of reasons, many of which are cultural and gendered). Malala is shown to be a 'fish out of water' in her Birmingham school; there are hints of her alienation and difficulty fitting in. Britain is not presented as the Promised Land for Malala. Through the relational lives of her mother, father and brothers (who play important roles in the documentary), Malala's heritage is emphasised. She will not simply be assimilated into British culture.

Conclusion

Each of the four texts in Malala's life narrative archive has involved mediation. This mediation has taken various forms whether transcription and translation, edited interview, collaborative writing, or reworking a text for a particular readership. These textual relationships reveal something significant about the transits of contemporary life narrative – the sorts of collaborations and mediations that seem necessary or inevitable, and those that are sanctioned by the cultures in which life narrative circulate. As critics, it is important that we do not default to reading such mediations as always problematic: for instance, as creating unequal power relationships, reinforcing stereotypes, and perpetuating the values of Western publishing industries and readers. Such processes are assumed to create inferior cultural texts. This may be common, but of course this is not what always happens, and it is troubling that such perceptions more often affect young writers and writers whose first language is not English.

In the examples discussed here, the processes of mediation have allowed Malala's narrative to reach different audiences across the globe via diverse genres (through news media, a mainstream memoir, a young-adult memoir and the most recent documentary, which received a cinema release and a wide international television release). As Khoja-Moolji contends, 'Across the archives, Malala emerges as an assemblage of positions and affects. She makes different kinds of knowledges possible in relation to the collectivities of Muslim women and men.' (552) These texts allowed Malala to address different questions about her life and her politics – layering knowledge and actively responding to criticisms about her, particularly her political agency and authorship of the life narrative texts. For example, as Mentzell Ryder notes, Malala's telling of the story of her shooting over multiple media and textual platforms shows how Malala 'takes control of the narrative about her attack':

> In Malala's story, the Taliban do not get the upper hand. Malala repeatedly explains that she had thought a lot about the possibility of such a moment – she had anticipated the attack and had decided what her action would be. Even though she was not able to carry out her plan at the time – they shot her before she could speak – she tells the story at every opportunity; she gives herself an active role, which she wants her audience to adopt. (179)

Further, as Mentzell Ryder notes, Malala works to authorise her voice by situating her narrative and political views within broader ideological contexts:

> By linking her philosophy to world religions, international leaders, Pakistani and Pashtun history and her family, Malala dispels any sense that her approach is merely a child's naiveté, or that her approach is unique to her. She calls upon a long international history of nonviolent action to refute calls for revenge. (180)

In creating a life narrative archive – through which her texts build, remediate and speak back to each other – Malala is able to further authorise herself and continue to circulate her ever-developing life narrative.

Collaborative life narrative archives show the potential power of life narrative as they reveal that life narratives do not have a natural end: stories can continue to develop and life narrative can engage in an ongoing dialogue with those responding to a text (for instance, critics). The mediation of Malala reveals the productive genre-crossing work that life narrative so often engages in. At a time when it is easy to be sceptical about the limits of life narrative texts, Malala's archive reminds us of the affordances of life narrative that lie within and beyond the critiques.

Notes

1. The film achieved critical acclaim and was translated into 45 languages and shown in 171 countries (Busch).
2. Theories of mediation, remediation and adaptation are also potentially relevant to this discussion and to the growth of adaptation within life narrative genres – whether book-to-film or adult text-to-young adult text. It is not within the scope of this article to explore these theories in any depth as the focus here is more on mediation as it is relevant to collaborative life narrative texts, and I use existing life writing theory to worth through these issues.
3. Anonymity and pseudonymity allow for the concealment of an identity and the protection of the authorial subject. But they also add a level of mystery, intrigue and even gimmickry:

enticing a readership that might not otherwise be interested. Pseudonymity may also signal the probability of raw honesty from the writer, and this is certainly how this blog is structured and presented to a readership: the blogger will provide something distinct from existing journalistic discourse. It will be reportage, but it will be direct and experiential: from a first-person witness. Indeed, this combination of narrative techniques (first person pseudonymous reportage of everyday life) and technology (the blog) had already proven to be a highly effective means of increasing the visibility of experiences 'on the ground' in the Middle East in the case of Salam Pax's blog *Where is Raed*? during the second Gulf War in 2003 (Whitlock 2007). The subsequent *A Gay Girl in Damascus* hoax, made during the Arab Spring in 2010, also attempted to utilise the wide-appeal of life writing from the Middle East to further the cause of the Syrian uprising.

4. At the time Malala lived in the Swat Valley in North-west Pakistan. The region was Taliban-occupied and the Taliban was trying to ban girls from attending educational institutions. Malala's family was actively involved in education; they ran a school in the region (Ellick and Ashraf).
5. The blog ran from January to March 2009 and was very popular with readers 'including Pakistani readers in the United Arab Emirates, India, the U.S., Canada and the U.K ... As well as being translated into English for the BBC, her entries were regularly reproduced in local Pakistani media.' (van Gilder Cooke).
6. According to Besharat Peer, Malala's community work and rise to prominence continued over the following two years:

> Malala became a celebrity in Pakistan in October, 2011, when Desmond Tutu announced her nomination for an international children's prize. It seems to have been the first time that her identity as the writer of the BBC diary became known to the broader public; the citation for her nomination mentioned her use of 'international media to let the world know girls should also have the right to go to school.'- Her public profile rose further after the Pakistan government awarded her the first National Peace Prize, in December 2011. 'In a situation where a lifelong school break was being imposed upon us by the terrorists, rising up against that became very important, essential,' she told a Pakistani television network.

7. Baig writes:

> There is no justifying the brutal actions of the Taliban or the denial of the universal right to education, however there is a deeper more historic narrative that is taking place here.This is a story of a native girl being saved by the white man. Flown to the UK, the Western world can feel good about itself as they save the native woman from the savage men of her home nation. It is a historic racist narrative that has been institutionalised. Journalists and politicians were falling over themselves to report and comment on the case. The story of an innocent brown child that was shot by savages for demanding an education and along comes the knight in shining armour to save her.The actions of the West, the bombings, the occupations the wars all seem justified now, 'see, we told you, this is why we intervene to save the natives.'

Disclosure statement

No potential conflict of interest was reported by the author.

References

Anonymous. "Diary of a Pakistani School Girl." *BBC news* 19 Jan. 2009. Web. 27 Aug. 2014. <http://news.bbc.co.uk/2/hi/south_asia/7834402.stm>.

Baig, Assed. "Malala Yousafzai and the White Saviour Complex." *The Huffington Post* 15 July 2013. Web. 20 Aug. 2015. <http://www.huffingtonpost.co.uk/assed-baig/malala-yousafzai-white-saviour_b_3592165.html>.

Bhutto, Fatima. Rev. of *I Am Malala*, by Malala Yousafzai with Christina Lamb. *The Guardian* 30 Oct. 2013. Web. 22 Oct. 2016. <https://www.theguardian.com/books/2013/oct/30/malala-yousafzai-fatima-bhutto-review>.

Brien, Donna Lee. "What about Young Adult Non-fiction?: Profiling the Young Adult Memoir." *TEXT* 32 (2015). Web. <http://www.textjournal.com.au/speciss/issue32/Brien.pdf>.

Busch, Anita. "National Geographic to Air Docu 'He Named Me Malala' Joining with Fox Searchlight." *Deadline.com* 18 June 2015. Web. 5 Oct. 2016.

Cellan-Jones, Rory. "The Banning of the Blog." *BBC* (Technology) 15 June 2012. Web. 27 Aug. 2015. <http://www.bbc.com/news/technology-18455348>.

ChristinaLamb.net. 17 Oct. 2016. Web.

Chowdhury, Afsan. "Malala Shooting: Does the BBC Share the Blame Too?" *BDNews24* 17 Oct. 2012. Web. 11 Sept. 2015. <http://opinion.bdnews24.com/2012/10/17/malala-shooting-does-the-bbc-share-the-blame-too/>.

Douglas, Kate, and Anna Poletti. *Life Narratives and Youth Culture: Representation, Agency and Participation*. Basingstoke: Palgrave, 2016.

Ellick, Adam B., and Irfan Ashraf. "*Class Dismissed*: Malala's Story." 2009. (Documentary). *The New York Times* 11 Oct. 2012. Web. 27 Aug. 2015. <http://www.nytimes.com/video/world/asia/100000001835296/class-dismissed.html>.

Ghafour, Hamida. "Malala Yousafzai: Backlash against Pakistani Teen Activist Spreads in Her Homeland." *The Toronto Star* 19 June 2013. Web. 20 Aug. 2015. <http://www.thestar.com/news/world/2013/07/19/malala_yousafzai_backlash_against_pakistani_teen_activist_spreads_in_her_homeland.html>.

He Named Me Malala. Dir. David Guggenheim. Imagenation Abu Dhabi FZ Participant Media, 2015. Film.

Jensen, Meg, and Margaretta Jolly. "Introduction." *We Shall Bear Witness: Life Narratives and Human Rights*. Madison: The University of Wisconsin Press, 2014.

Khoja-Moolji, Shenila. "Reading Malala: (De)(Re) Territorialization of Muslim Collectivities." *Comparative Studies of South Asia, Africa and the Middle East* 35.3 (2015): 539–556. Print.

Kurz, Katja. *Narrating Contested Lives: The Aesthetics of Life Writing in Human Rights Campaigns*. Heidelberg: Universitätsverlag Winter, 2015. Print.

Makai, Gul. (Malala Yousafzai). Original Blog Entries. *BBC Urdu* 9 Jan 2009. Web. 11 Sept. 2015. <http://www.bbc.com/urdu/pakistan/story/2009/01/090109_diary_swatgirl_part1.shtml>.

Mentzell-Ryder, Phyllis. "Beyond Critique: Global Activism and the Case of Malala Yousafzai." *LiCS* 3.1 (2015): 175–187. Print.

Peer, Basharat. "The Girl Who Wanted to Go to School." *The New Yorker* 10 Oct. 2012. Web. 27 Aug. 2015. <http://www.newyorker.com/news/news-desk/the-girl-who-wanted-to-go-to-school>.

"I Am Malala: Comparing the Young Reader Edition to the 'Original'. *Ph.Ds and Pigtails*. Blog. Web. <https://phdsandpigtails.com/2015/03/12/i-am-malala-comparing-the-young-reader-edition-to-the-original/>.

Rev. of *I am Malala* by Malala Yousafzai and Patricia McCormick (Young Readers Edition). *Kirkus Reviews* 19 Aug. 2014. Web. 14 Oct. 2016. <https://www.kirkusreviews.com/book-reviews/malala-yousafzai/i-am-malala/>.

Schaffer, Kay, and Sidonie Smith. "E-witnessing in the Digital Age." *We Shall Bear Witness: Life Narratives and Human Rights*. Eds. Meg Jensen and Margaretta Jolly. Madison: The University of Wisconsin Press, 2014. 223–237. Print.

Tahir, Madiha R. "Reading Malala's Diary." *Tanqueed: A Magazine of Politics and Culture* 12 Nov. 2012. Web. 11 Sept. 2015. <http://www.tanqeed.org/2012/11/reading-malalas-diary/>.

van Gilder Cooke, Sonia. "Pakistani Heroine: How Malala Yousafzai Emerged from Anonymity". *Time World* 23 Oct.ober 2012. Web. 27 Aug. 2015.

Westhead, Rick. "Brave Defiance in Pakistan's Swat Valley." *Toronto Star* 26 Oct. 2009. Web. 27 Aug. 2015. <http://www.thestar.com/news/world/2009/10/26/brave_defiance_in_pakistans_swat_valley.html>.

"Young Journalist Inspires Fellow Students." *Institute for War and Peace Reporting* 5 Dec. 2009. Web. 27 Aug. 2015. <https://www.iwpr.net/global-voices/young-journalist-inspires-fellow-students>.

Yousafzai, Malala, with Christina Lamb. *I am Malala: The Girl Who Stood Up for Education and Was Shot by the Taliban*. New York: Little Brown, 2013. Print.

Yousafzai, Malala, with Patricia McCormick. *I am Malala: The Girl Who Stood Up for Education and Was Shot by the Taliban (Young Readers' Edition)*. New York: Little Brown, 2016. Print.

Yousafzai, Malala. "Moving Moments from Malala's BBC Diary." 10 Oct. 2014. Web. 27 Aug. 2015.

Yusuf, Huma. "About the Malala Backlash." *The New York Times* 18 July 2013. Web. 20 Aug. 2015. <http://latitude.blogs.nytimes.com/2013/07/18/the-malala-backlash/?_r=1>.

Yusufza, Mushtaq. "Pakistani Teen Blogger Shot by Taliban 'Critical' after Surgery." *NBC News* 9 Oct. 2012. Web. 20 Aug. 2015. <http://worldnews.nbcnews.com/_news/2012/10/10/14332088-pakistani-teen-blogger-shot-by-taliban-critical-after-surgery?lite>.

Remembering Violence in Alice Pung's *Her Father's Daughter*: The Postmemoir and Diasporisation

Anne Brewster

ABSTRACT

Alice Pung's postmemoir of the after-effects of political violence maps a discursive trajectory from (1) her father's survivor memory of the Cambodian genocide, to (2) her own postmemory as a second-generation Asian-Australian, to (3) the latter's remediation as social memory within the Australian (trans)national imaginary. Hirsch describes the family as 'the privileged site of the memorial transmission' of trauma. In *Her Father's Daughter*, Pung parallels the heroic narrative of her father's survival of 'a real and bloody social revolution' (*HFD*, 48) with the more modest narrative of her own embodied travails with 'authentic feeling' (21) regarding her affective connectivity with her extended family and the cultural and geographical landscapes they inhabited. Her postmemorial journey is one into her own heart, variously described as 'a deformed dumpling' (28) and 'rotting fruit' (32). Literary texts such as Pung's can bring about the timely reanimation of the post-settler state's archives through investing them with familial forms of mediation and aesthetic expression. In *Her Father's Daughter*, disaporic subjectivity is articulated through the mapping of transnational and transgenerational histories.

This article examines how the Anglophone Chinese-Cambodian Australian writer Alice Pung uses the genre of the memoir to reflect on extreme political events and their transgenerational effects within the family, articulating in the process the emergence of diasporic subjectivity. It demonstrates how the relatively new body of minority life writing is extending the parameters of the field of life writing by reconfiguring our understanding of transnational memory and historicity within the white nation. Pung's memoir, *Her Father's Daughter*, addresses the issue of how, by commemorating violent histories, minority life writing undertakes significant (non-governmental) political, cultural and aesthetic work, contributing to the formation of diasporic imaginaries and the creation of minority groups as a public. The book relates the story of Pung's family's migration from war-torn Cambodia to Australian suburbia, carrying on the project of her first memoir *Unpolished Gem* (2006); however, unlike *Unpolished Gem*, it focuses largely on the narrator's relationship with her father and it includes the story of her family's life in Cambodia before and during the Pol Pot regime. Pung, who tells the story of her father's escape from Cambodia and his subsequent life as a businessman and father in

Australia, is a second-generation survivor of genocide. In reflecting on the multiplication of sub-categories of the term 'life writing', I borrow from Marianne Hirsch and Leslie Morris to define *Her Father's Daughter* as a postmemoir and, further, from Thomas Couser and Ruth Charnock to argue that in this book the broad genre of life writing hybridises and bifurcates into two intertwined narratives, a patriography[1] (Couser) and a filiography[2] (Charnock 56). This discussion starts from the assumption that family memoirs are relational, foregrounding the incomplete process of the division of the 'in-dividual' (Stocks 81) from her parents, and the persistence of the parental experience within the self. This fact is particularly evident in families that have experienced catastrophic events, the impact of which reverberates down the generations.

The term 'patriography' was coined by Thomas Couser in 2009 and, as a new genre, has been little theorised, despite a special issue of *Life Writing* dedicated in 2014 to the genre and Stephen Mansfield's important book, *Australian Patriography: How Sons Write Fathers in Contemporary Life Writing*. As Mansfield's title suggests, his study focuses solely on men writing about their fathers. Indeed, Couser argues that the majority of memoirs about parents are written by men about their fathers (21) and that many describe attempts by the writer to fashion a relationship with a father who is either absent or emotionally reserved. These memoirs focus on 'paternal secrecy and filial sleuthing' (Couser 22) and constitute a narrative where 'an adult child explores something hidden, withheld or misrepresented in the father's life' (Couser 22). These memoirs work both to memorialise the writer's father and to repair the relationship with his father. To some extent *Her Father's Daughter* mirrors this concern with secrecy and sleuthing. It also raises broader questions that underlie the commemoration of violent political histories, namely *whether* to remember violence, *how* to remember it and *why* it is to be remembered (Landres and Stier 8). These questions point to the fact that there are often contests over the act or the meaning of remembering political violence. In what follows I examine memory as a generational contest and analyse Pung's own diasporic investments in the memorialisation of the Cambodian genocide as a second-generation Asian-Australian.

While one reviewer of *Her Father's Daughter* suggests that 'the father is at the forefront of this work' (Walker 64), another, Thuy On, suggests that Pung's father 'shares equal billing with the author' (24). He describes the book as a 'filial love song' (On 24). If *Her Father's Daughter* is a loving tribute to her father, it performs its tribute through the act of witnessing, which Pung undertakes as an ethical response to catastrophe. In some sense, the act of witnessing authorises Pung's exposure of intimate details of her father's life. In relational memoirs, as Couser suggests, the focus often shifts from parent to child, and in *Her Father's Daughter* Pung's act of witnessing doubles to produce two stories – a patriography which revolves around Pung's father, Kuan Kieu Pung, and his experience as a survivor of the Cambodian genocide, and a filiography which narrates Alice Pung's recovery from transgenerational trauma and her search for independence and love. Both the patriography and the filiography aim to fashion an ethical mode of postmemorial intersubjectivity. Pung writes both stories in the third person with chapters focalised alternately through her father's and her own narrative point of view. In the interplay between the patriographic and the filiographic narratives, as the focus shifts from parent to child, we see the emergence of a second-generation diasporic imaginary.

Bearing witness to catastrophe entails the narration of embodied affects. *Her Father's Daughter* figures the affective bodily impact of the intergenerational transfer of memory upon Pung as the descendent of a survivor. If her father represents those who lived through the Cambodian Killing Fields, Pung is a member of the second-generation who internalise, through what Marianne Hirsch describes as postmemory, the bodily and psychic after-effects of the trauma their parents experienced. While postmemory, Hirsch suggests, is not identical to memory, it approximates memory and shares its affective force. She argues that within the families of Holocaust survivors, survivors often express not only narrative memories but also 'a chaos of emotion' associated with their experience of trauma (Hirsch 111). When he arrives in Australia, Pung's father, Pung says, had been reduced psychically and materially to the condition of 'human debris' (73) and the life he reconstructs for himself is marked by anxiety, paranoia (192), bad dreams and 'incessant worrying' (62). Pung attempts to plumb the silence and what she terms the 'dismemory' (191) that shroud his past. 'Dismemory' is a word she coins to describe a deliberate and conscious act of forgetting (distinguishing it from the *unconscious* act of repression).[3]

The book tells us that Pung's father did talk about some of his experiences under the Khmer Rouge throughout her childhood and that he used humour to counteract the trauma of those experiences (Pung, 'Interview'). However, in order to protect them, 'he never wanted his children to know' (109) the full details of his own family's experience of suffering and horror. We are told that '[t]o live a happy life, he believes, you need a healthy short-term memory, a slate that can be wiped clean every morning' (5). In this way he intended that 'all his feelings would be only a day old' (211). Pung's father's way of dealing with disturbing memories had been to not talk about them, believing that '[t]alk led to nothing and nowhere' (197). Like many of his generation in Australia, his desire is for 'privacy' (223) and '[m]inding [his] own business' (197), which affords him 'a kind of peace' (223). He resists the therapeutic discourse of psychoanalysis; in a chapter written in his voice, Pung ventriloquises her father, saying that the practice of treating post-traumatic stress disorder is 'rubbish' and that it 'stop[s] a person from moving on in life' (197).

Numerous commentators on trauma and violence have reminded us of the cultural specificity of the hegemonic Western psychological model and the premise that speaking about trauma is the most effective way to master it (Craps 23; Argenti and Schramm). They have researched various post-conflict global cultures where people have chosen to live with memories of violence by not talking about them. This includes people in Buddhist cultures who have a different attitude to suffering. Whereas Pung understands suffering in terms of Western notions of subjectivity and personhood – where interpersonal relationality and healing are consolidated through speech – her father and other members of his family, like many other Cambodians of their generation, find a Buddhist worldview a source of resilience and emotional strength (Kidron 217). In Buddhist cultures, silence and acceptance in the face of suffering is an index of strength rather than weakness or insufficiency, and it is considered respectful of the memory of suffering and of those who have died (Kidron 211).

It is also empowering in cases where the victims of political violence continue to live alongside its perpetrators (Argenti and Schramm 16). We see several examples of people living in this kind of proximity in *Her Father's Daughter*. Pung describes with surprise her father and her uncle, on a trip to the country, talking 'calmly and casually, as if he were some ordinary neighbour' (216), with an old man who had once headed the children's army in their collective. We had read earlier in the book of terrible acts this man had committed against children in the children's army. The narrator comments at this point upon the 'goldfish memory spans' (216) – the ability of people involved on opposite sides of deadly violence to employ 'dismemory' in order to be able to live together. Although Pung describes this ability to bear oneself through painful memories as a form of 'grace' (216), her use of the image of the 'goldfish memory spans' in the same sentence qualifies the reverential tone and indicates her difficulty in understanding and accepting the compromises and the forgetting undertaken by people in these circumstances. Another example of this kind of collective dismemory that Pung observes in Cambodia is when she notes ironically that the bodyguards her uncle employs to safeguard his family are ex-soldiers (from the Khmer Rouge regime) (215). These may have been the very people that terrorised his family during the regime. We can extrapolate further from the high security that her uncle needs to protect his family that Cambodia has exchanged one form of violence for another (although the memoir itself does not pursue this issue).

Pung's father's disinclination to participate in a therapeutic discourse about the violence of the Pol Pot regime is echoed by another survivor, his sister-in-law Suhong, who says to Pung, while the latter is in Cambodia undertaking research for her book, '[a]h, there is nothing to say about those bad times … [t]hinking about them only makes you feel sad all over again' (202). Her aunt Suhong's injunction to not talk about that period may well index the Cambodian national imaginary of her generation. In an interview, Pung relates that there are very few apparent material traces (apart from bones) of the Khmer Rouge regime in present-day Cambodia; for example, the huts in which people lived have been overgrown by forest. She extrapolates from this that many Cambodians prefer to let present exigencies 'grow over' their genocidal history. She relates that her friends in Cambodia say that they don't know anything about that 'bad time' and that their parents don't talk to them about it (Themonthlyvideo).

The disjuncture between her father's silence and Pung's imperative to tell the story of the Cambodian genocide is mediated by the intertwining of the patriographic and filiographic narratives and reflected in the aesthetics of the postmemoir. Her narration of her father's experience of the Cambodian genocide is summative and documentary, seeking in part to reconstruct historical events in order to redress Pung's lack of knowledge about her father's past in order to memorialise his life, but also to understand him and her childhood better. The postmemoir represents, to borrow Eva Hoffman's words, 'a deeply internalised but strangely unknown past' (quoted in Hirsch 108). Pung relates that she did not start the postmemoir with the Cambodian section, instead positioning it half way through the book so that it would not be too confronting to readers (193). This narrative delay in effect mimics the sense of belatedness that the revelation of her father's history produced for Pung. It is this belatedness which motivates the quest narrative of the postmemoir.

In the patriographic narrative, Pung commemorates her father's resourcefulness and courage, both during the four years of the Pol Pot regime, and then the following 30 years in Australia during which time he moved from being a factory worker to a Retravision franchisee with substantial investments in real estate. The epilogue of the book celebrates his 60th birthday. If, in the words of Thuy On, *Her Father's Daughter* is a 'filial love song', we can see that for Kuan Kieu Pung the family unit has been an important means by which he has rebuilt his life and that it has provided a powerful unit of cohesion in response to loss and mobility. If the memoir constitutes a patriographic tribute to her father, this tribute also takes the form of a filiography, as suggested by the book's title. At stake in the telling of family history through the patriographic and filiographic narratives is the commemoration of Pung's father's traumatic past. However, *Her Father's Daughter* is also centrally concerned with the nature of the intergenerational transfer and its impact on Pung as a young woman. While Pung's narration of her father's story of survival is one of heroic masculinity, the gendered narrative about her own life is a comparatively modest and familial narrative about the 'small things' of the everyday (5), positioned in domestic space and the affective life of the body. If, for Pung, the family is the privileged site of memorial transmission (Hirsch 110), this is because diasporic Chinese, as she tells us, are 'destined never to feel a sense of belonging; knowing they would never be *a part* unless they kept themselves *apart* and hid what was most important of their heritage inside the home' (194). The private space of the home is a repository of survivor memory and the embodied generational aftereffects of political violence.

Marianne Hirsch writes of second-generation children of Holocaust survivors that:

> To grow up with such overwhelming inherited memories, to be dominated by narratives that preceded one's birth or one's consciousness, is to risk having one's own stories and experiences displaced, even evacuated, by those of a previous generation. (107)

Her Father's Daughter does indeed portray Pung's struggle to recognise and fulfil her own needs. It carries on the story of Pung's difficulty with love which is so poignantly protrayed in her earlier memoir, *Unpolished Gem*. While always compassionate and respectful, she describes her father's paternalistic control in *Her Father's Daughter* as claustrophobic: '[t]here were too many attachments in this world, she realised, and sometimes love bound too tightly' (44). Chafing against paternal control leads to feelings of guilt (42; 93) and the sense that Pung cannot approximate her parents' suffering or their intensity of life experience of living through 'a real and bloody socialist revolution' (48). When she seeks to establish an intimate relationship with Teodore, she has trouble individuating herself from their experience of adventure and risk early in their marriage:

> [h]er parents had spent their honeymoon in the jungles along the Thai-Cambodian border, fleeing from the Khmer Rouge. For true intensity of experience you could not beat that. (37)

She compares their experience with her own inability to take risks in love (she is 'too careful' [228]) and to invest in and sustain an intimate relationship. This leads to feelings of melancholy and sadness (88) as she struggles with abjection and anxiety. If her father's struggle is one to survive the rigours of the killing fields, Pung's antagonist seems to be her own heart which she describes variously as '[t]he deformed dumpling' (28) and 'rotting fruit' (32). In an interview she defines 'Dismemory' as 'a memory that has been amputated like a phantom limb; you can feel it but it is not here any more. This is what effects the second generation' (Themonthlyvideo). Pung is describing here the embodied process of re-membering her own body which often seems a stranger to her, recalcitrant and alarming. She negotiates the corporeal effects of her father's fear and anxiety and his surveillance and cloistering (65) of her femininity.[4] If her father has decided not to 'wast[e] energy on emotions' (198), Pung struggles to acknowledge and honour her feelings. The body is after all the site where power impacts and where race and gender are regulated. In *Her Father's Daughter*, we observe how gender insects with Chineseness and how Pung resists and negotiates her father's concept of ideal South-east Asian femininity which he has monitored since her childhood.[5] The filiographic narrative of *Her Father's Daughter* constitutes a tribute to her father and an affirmation of her affiliation with him, but also a recovery narrative where she avers that she must in some measure disaffiliate from him in order to move towards building – both materially and psychically – what she describes in the closing pages of the postmemoir as 'her future family home' (233).

The postmemoir thus constitutes two stories – both the commemorative patriography and a filiographic narrative – which negotiate what Hirsch calls 'overwhelming inherited memories' of catastrophe (5). The latter relates how the individual takes the initiative to live a meaningful life in the face of such a family history in a way that performs care both for the beloved father and the self. The postmemoir can be seen as a 'laboratory in which selves are made and re-made in reflective and contested' ways (Givoni 20). Part of this remaking of the self involves a search by the filiographic subject for what one reviewer calls 'her ethnic and cultural roots' (Oliver 32). I have suggested that the history of her family's experience of the Cambodian Genocide has a very different meaning for her father and his generation than it has for Pung. She feels that by not sharing his family history with her, her father had withheld a large part of her heritage. Concomitantly, the recovery of this heritage is a powerful technology of the self for Pung, and writing *Her Father's Daughter* represents a significant subjectifying event.

In an interview on *The Book Show*, Alice Pung said, 'I finished *Her Father's Daughter* when I was 30 which was the age my father was when he came out of the killing fields.' Elsewhere, she comments that the book 'was a difficult book to write'.[6] These remarks suggest that the project of writing *Her Father's Daughter* was an important personal achievement and that it was a milestone for her in the same way that her father's survival of and escape from the Cambodian Killing Fields marked a *rite-de-passage* which allowed him to be 'born again' (229) and start a new life. By bringing the two milestones of turning 30 into proximity and aligning her and her father's stories, Pung suggests that the recovery of her family history and its remediation as literature provided the emergence of a new diasporic identity for her. In jointly telling her father's story of survival and migration and her own story of negotiating transgenerational trauma as a member of a minoritised group within the white nation, Pung in effect tells us a story of what it meant to 'grow up Asian in Australia'.[7] (Post)memorial work such as *Her Father's Daughter* has emerged, I would suggest, at a historical juncture when the majority of the heterogenous group we could describe as 'Asian Australians'[8] are first- or second-generation Australians (Ommundsen 507), and reflects a generational interplay of forgetting and remembering.

Lily Cho argues that diaspora is not simply an act of migration but a condition of subjectivity. She says that one is not born diasporic but '*becomes* diasporic through a complex process of memory and emergence' [emphasis in the original] (Cho 21). The subject's relation to the past is crucial in this matter. Malissa Phung describes diasporisation as an 'ongoing process of discovering and mending an always tenuous relationship to the past' (2). In *Her Father's Daughter*, Pung negotiates her sense of belonging and home across a number of geo-national contexts. The book opens with Pung on a writer's residency in China, where she returns to the villages which are her 'ancestral hometown' (11) – the birthplaces of her father's parents. She is searching for 'an authentic feeling'

(21) of recognition or belonging, but little is familiar and she seems to identify with the label of '[f]oreign ghosts' (12) assigned to her by some local children. She relates that although her father had raised his children to be proud of their heritage of being Chinese and of 'having ancestry from the Middle Kingdom' (5), she feels a 'complete stranger' (Pung, Interview) in China. Her postmemories of China (derived from her grandmother's stories) pale in comparison with the significance of the place that Cambodia occupies in her psyche and embodied life. The story that she feels compelled to tell in *Her Father's Daughter* is not that of her Chinese heritage but of the Cambodian genocide. It is her father's silences about the trauma of Cambodia which are in fact definitive of her subjectivity. She says that he had:

> drilled into [his children] that they were part of a Chinese culture that spanned centuries, which was true; and made sure that they were also bonafide born-in-Australia kids. But in doing so, he had wiped out *the most significant part of their identity.* [emphasis added] (193)

Pung invests the transgenerational effects of the trauma of the Cambodian genocide with a central significance as the principal subjectifying force in her life. She experiences the absence of her father's memories as a plenitude. The narrative climax of the postmemoir occurs in the return to the site of her father's originary trauma, namely the field where her father had been forced to bury the dead. We see the aesthetics of postremembrance – the impulse to document, to recover a trace – playing out in the figuring of the field which exemplifies a duality: full/empty. Pung identifies the field according to its emptiness: 'there were not even bones left. None of these people seemed to have existed' (215). Yet, paradoxically, 'this flat stretch of nothing was everything … all her father's life had been filling this emptiness' (213–14). And, of course, the flat space of the field and its representational emptiness has also prompted the loquacious postmemorial narration of the self.

The putatively empty field is replete with public images and narrative. Leslie Morris reminds us of the impossibility of an unmediated relationship with the past (293), and when Pung writes that there 'were not even bones left', this inevitably recalls the widely circulating public images of skulls and bones, the archived bodily remains of the Cambodian Killing Fields. Morris notes that postmemoir is characterised by 'metareflection' about the originary traumatic event (293). Pung's choice of the trope of the empty field to signify her father's trauma suggests an awareness about the publically mediated nature of memory. It also ironically highlights Pung's ambivalence over the ways in which her father (dis)remembers the Cambodian genocide. The slightly hushed and lyrical minor epiphanies in the chapter titled 'the field (ii)' are moderated by the presence of ex-soldiers of the Khmer Rouge regime who flank the empty field in their new role as her family's bodyguards, and the amiable conversation her father and uncle have with the former head of their collective children's army. These troubled and troubling narratorial observations modulate the breathless tone of Pung's narrative at this point. The elegiac impulse of memorialisation is accordingly qualified by the statement that 'she would never understand' this country (217). This ambivalence aligns with Lily Cho's description of the diasporic subject as emerging dialectically 'from a history that is both restorative and incomplete' (17).

Pung's statement about not understanding Cambodia foregrounds her positionality as a second-generation Asian-Australian of Chinese-Cambodian descent who identifies not with the originary trauma of the Killing Fields but with its diasporic inter-generational and transnational after-effects. She avers that she doesn't fully comprehend the contradictory demands of the past in the present for the survivor generation (for example, the demand for retributive justice brushing up against the need for former perpetrators and victims to live together). Although she enjoys the pleasures of connectivity and seeks proximity with her extended family in Cambodia, it is clear that Pung's relationship with Cambodia of the past and of the present day is ambiguous; it is constituted by both continuity and rupture. Lily Cho further explains that diasporic subjects emerge in the process of 'turning back upon ... markers of the self – homeland, memory, loss – even as they turn ... away from them' (11).

Lily Cho further characterises diasporic subjects as emerging in relation to power, specifically power that marginalises or excludes, usually through racialisation. They are defined in relation to the norm of the nation state. The white Australian nation is one backdrop against which Pung narrates her postmemoir. Throughout the book, white Australians are seen as the default citizens – 'the Australians' – with whom minoritised people exist in a relationship of supplementarity. While *Her Father's Daughter* works to challenge the hegemony of the white nation's historicity and to establish a minoritised group as a public, Pung is nonetheless aware of the co-option of the diasporic Chinese as a 'model minority' (194) on account of their patriotism, their aspirations for modernity and upward mobility and their work ethic.[9] As a product of a middle-class Chinese-Cambodian Australian author, *Her Father's Daughter* has been seen as an index of this purported 'model minority', as the reviews of the book indicate. It has been described as the 'epitome of the migrant success story' (On 24), to be distinguished from 'a conventional migrant misery story' (Sullivan).

The aversion to the migrant 'misery story' and 'book[s] about refugees' (Sullivan) evinced by reviewers would appear to confirm Pung's father's assessment of white Australians as not wanting 'to read about too much suffering' (193). This disavowal of suffering indexes the white nation's belatedness in acknowledging and responding ethically to traumatic minority histories, and confirms the efficacy of Pung's decision to position the horrific story of the Pol Pot regime halfway through the novel rather than at its opening in order to make the book less challenging to a white readership. Pung is aware that majoritarian audiences are curious about minority histories and cultures; they:

> wanted to know what it was like to look through the windows of those concrete houses in Braybrook,[10] what it was like to open the fly-screen doors and see of sliver of the life inside. (90)

However, this curiosity exists alongside their disavowal and nescience of minority suffering. One reviewer's comment that the white Australians with whom Pung's father came into contact would have had 'no way of knowing what fearful events [had] brought him among them' (Walker 64) is noteworthy. Minoritised literatures such as *Her Father's Daughter* challenge the ignorance that the reviewer observes which, in the face of the wide media coverage of the atrocities of the Pol Pot regime, persists in shoring up the dominance and universality of whiteness.

Pung is, in effect, positioned in some of the reviews as the broker of the 'new migrant success story'. According to one reviewer, she occupies the role of double-agent, the insider who is both assimilated to majoritarian culture (as a member of the 'model minority') and an insider-informer of the impenetrable world of 'new migrants':

> Pung presents the experience of new migrants and her father's reaction to his new country with inside knowledge of his wonder and gratitude and relief. (Owen 106)

Pung herself identifies one of the characteristics of diasporic peoples transitioning into a 'model minority' as the acquisition by their children of English language competency (194). Indeed, English language competency enables Pung to tell her father's story to an Australian (and a global Anglophone) audience, yet there is a cost to this. Alice Healy comments on how, in the re-negotiation of diasporic Asian identities, 'this affective interplay between languages' (5) informs the remediation of transgenerational trauma and histories of violence as literature. In her discussion of Pung's postmemorial 'translation' of her parents' lives, Healy points to the role of English in the erasure of maternal memory (that of Pung's mother and grandmother)[11] (Healy 12).

In closing, I would like to return to the broader questions which underlie the commemoration of violent political histories that I mentioned at the beginning of this article, namely *whether* to remember violence, *how* to remember it and *why* it is to be remembered. I have argued that Pung's choice to remediate her family history as literature establishes Chinese-Cambodian Australians as a public within the white nation. However, this is a contested process, as Pung's postmemoir indicates, as father and daughter want to tell very different stories of diasporisation. While her diasporic subjectivity is fashioned by her postraumatic witnessing of the Pol Pot Regime, his desire is for a heroic story of 'glories of civilisation' which celebrates a heritage of the Middle Kingdom (5).

I have also argued that there are different stakes in memorialising the traumatic history of the Pot Pot Regime for people living in and outside Cambodia. For Pung, the writing of her postmemoir is an act of turning both backwards and forwards; it is a platform that enables her to establish transnational connectivities and also confirm her identity as a diasporic subject within the white Australian nation. Pung's postmemory is converted into multidirectional transnational social memory. Gillian Whitlock observes that testimony can 'move beyond the auspices of nation … and engage with other testimonial transactions' (152). On the one hand, *Her Father's Daughter* forges global networks and links her to various diasporic communities and literatures beyond the borders of the white nation. On the other hand, *Her Father's Daughter* and other minority life writing that focuses on issues of (post)memory prompt us to reconsider the post settler multicultural nation's historicity. That historicity is essentially transnational and multidirectional. For example, majoritarian recognition of minority people's cultural trauma issuing from war and genocide, I'd suggest, provides the occasion for a belated and unconscious remembering of the trauma of Australia's postcolonial state-sponsored history of violence which, for many majoritarian Australians, remains denied or disavowed. Within the transnation, settler history is brought into proximity and resonates with other violent political histories. Concomitantly, for minoritised diasporic constituencies, their arrival in the post-settler nation is 'an arrival within a history no longer simply their own' (Caruth 188), and their own histories open up into dialogue, both vertically with the majoritarian white nation and horizontally with other minority and indigenous histories.

There are connectivities between minority life and testimonial writing, and Aboriginal writing in the same genres, which address issues relating to histories of violence and differential citizenship. Just as Pung describes the family as providing a refuge for hiding 'what was most important of their heritage inside the home' (194), for example, I have argued that much Aboriginal women's life writing portrays the family as a gendered site for the resistance to assimilation (Brewster). One such text is Sally Morgan's postmemoir *My Place* (1987), which, like much other Aboriginal women's life writing, testifies to the survival and the traumatic after-effects of genocide. In Morgan's paradigmatic text her grandmother's silence regarding her own paternity and that of her daughter and Morgan's efforts to uncover this originary violence recalls Pung's project of researching her father's experience of the Killing Fields in *Her Father's Daughter*. Further, the Wiradjuri writer Anita Heiss has attested to the horizontal connectivities between indigenous and minoritised writers. She acknowledges Alice Pung's influence in the conceiving of Heiss' memoir *Am I Black Enough for You?* (2011). Heiss describes being inspired by Pung's iconoclasm and her engagements with racialising stereotypes in her life writing.[12]

However, to say that a majoritarian recognition and incorporation of minority genocidal histories into the national archival imaginary prompts the recall of that other occluded genocide – of the first nations people – is not to assert any easy equivalence between the cultural and political histories of indigenous and minority people in Australia. Issues of class,[13] religion, sovereignty, land and formations of whiteness and coloniality differentially impact upon and inflect the subjection and racialisation of these two constituencies. The important contribution that both minority and Indigenous life writing make is to redefine the historicity of the white nation. Literature from minoritised authors transforms these constituencies from being merely 'watchers'[14] to being the *writers* of transnational histories.

Notes

1. My thanks to Gillian Whitlock for drawing my attention to this category of patriography.
2. This term, first used by Charnock, can be seen as deriving from both the Latin terms *filius* (son) and *filia* (daughter) and can therefore, I suggest, refer to either a male or a female biography.
3. Pung, Interview by Anita Barraud.
4. Her father's is a story of heroic masculinity; a man who comes to Australia with his wife, sister and mother whom it is his job to protect (this includes later managing the femininity of his daughters). One might argue that his masculinity is subtended by a degree of cloistering of the women of his immediate family (his wife and daughters).
5. See for example, her father's ideal of South-east Asian girls as innocent and 'tender morsels' (102). This theme is explored more fully in *Unpolished Gem*, although in the earlier memoir most of the surveillance of Pung's ideal Asian femininity is conducted by her mother, who herself has internalised a sense of the inferiority of her own Asian femininity and of the superiority of idealised white femininity (see *Unpolished Gem*, 242).
6. Pung, "No Subject Line."
7. This is the title of the book that she edited in 2008.
8. Apart from a small number descended from Chinese who immigrated during the gold rush days, for example.
9. She observes throughout the postmemoir that her father exemplifies these characteristics (see for example, 193–7).
10. This is the suburb where Pung grew up. It was formerly a housing commission area and is described on ABC news as 'one of the ten most disadvantaged suburbs' in the state of

Victoria. See "Life in Victoria's 'Bronx': Braybrook, Where 'Every Second House was Dealing Drugs'."

11. Healy is referring here to Pung's first book, the memoir *Unpolished Gem*.
12. On her blog, Anita Heiss explicitly references *Unpolished Gem*. https://anitaheiss.wordpress.com/2014/05/29/it-all-starts-with-a-conversation/.

    > Many of you will know of my memoir on identity published by Random House in 2011. The work was originally inspired by Alice Pung's *Unpolished Gem*. It was the first line of Alice's book 'This story does not begin on a boat,' and the later line 'There are no Wild Swans or Falling Leaves' that struck me immediately. I knew *I* had to write a story about Aboriginal Australia that *didn't* begin in the desert, and *did not* have didgeridoos playing in the background, and there would be no dot paintings to be seen. (Accessed 1 August 2016).

    I am indebted to Imogen Mathew for bringing this to my attention.
13. Pung's father was from the elite business class of Phnom Penh (125). He spoke English and French in addition to Khmer and Chinese (117). The privileges of this class location would have impacted upon his mobility and success in Australia. I would argue that this success might have enabled Pung to tell the other painful and shocking story of his and his family's loss under the Khmer Rouge.
14. The term is taken from *Unpolished Gem* (186), where Pung describes the tacit social exclusion of minoritised students and their parents through their relegation to the role of 'watchers' at her school valedictory evening. For a more explicit description of how racialised exclusion plays out in a school environment, see Pung's young adult novel Laurinda (2014).

Acknowledgements

My thanks to David McCooey and Maria Takolander for inviting me to present an earlier version of this article at the Life Writing symposium at Deakin University in February 2016. Thanks are also due to Rose Arong for her excellent research assistance and the two anonymous reviewers for their feedback.

Disclosure statement

No potential conflict of interest was reported by the author.

Funding

This work was supported by the Australian Research Council [DP140100552].

References

Argenti, Nicolas, and Katharina Schramm. "Introduction." *Remembering Violence: Anthropological Perspectives on Intergenerational Transmission*. Eds. Nicolas Argenti and Katharina Schramm. New York and Oxford: Berghahn Books, 2010. 1–39. Print.

Brewster, Anne. *Aboriginal Women's Autobiography*. Sydney: Oxford University Press in association with Sydney University Press, [1996] 2016. Print.

Caruth, Cathy. "Unclaimed Experience: Trauma and the Possibility of History." *Yale French Studies 79 Literature and the Ethical Question* (1991): 181–92. Web. 1 Feb. 2016.

Charnock, Ruth. "Incest in the 1990s: Reading Anaïs Nin's 'Father Story'." *Life Writing* 11.1 (2014): 55–68. Taylor & Francis Online. Web. 30 May. 2016.

Cho, Lily. "The Turn to Diaspora." *Topia* 17 (2007): 11–30. Print.

Couser, Thomas. "Paper Orphans: Writers' Children Write their Lives." *Life Writing* 11.1 (2014): 21–37. Taylor & Francis Online. Web. 19 Feb. 2016.

Couser, Thomas. "In My Father's Closet: Reflections of a Critic Turned Life Writer." *Literary Compass* 8.12 (2011): 890–99. Print.

Craps, Stef. *Postcolonial Witnessing: Trauma Out of Bounds*. Houndmills: Palgrave Macmillan, 2013. Print.

Givoni, Michal. "The Ethics of Witnessing and the Politics of the Governed." American Political Science Association Annual Meeting, Seattle, 1–4 Sept. 2011. Web. 1 Jun. 2016.

Healy, Alice. "Unable to Think in My Mother's Tongue': Immigrant Daughters in Alice Pung's *Unpolished Gem* and Hsu-Ming Teo's *Behind the Moon*." *Transnational Literature* 2.2 (2010): 1–13. Web. 30 Jan. 2016.

Hirsch, Marianne. "The Generation of Postmemory." *Poetics Today* 29.1 (2008): 103–28. Duke University Press Journals Online. Web. 1 Feb. 2016.

Kidron, Carol. "Silent Legacies of Trauma: A Comparative Study of Cambodian Canadian and Israeli Holocaust Trauma Descendant Memory Work." *Remembering Violence: Anthropological Perspectives on Intergenerational Transmission*. Eds. Nicolas Argenti and Katharina Schramm. New York and Oxford: Berghahn Books, 2010. 193–228. Print.

Landres, J. Shawn and Oren Baruch Stier. "Introduction." *Religion, Violence, Memory, and Place*. Eds. Oren Baruch Stier and J. Shawn Landres. Bloomington & Indianapolis: Indiana University Press, 2006. 1–12. Print.

"Life in Victoria's 'Bronx': Braybrook, Where 'Every Second House was Dealing Drugs'." *7:30*. Jose Taylor. Australian Broadcasting Corporation. 20 Jul. 2015. Web. 12 Feb. 2016.

Mansfield, Stephen. *Australian Patriography: How Sons Write Fathers in Contemporary Life Writing*. London: Anthem Press, 2013. Print.

Morris, Leslie. "Postmemory, Postmemoir." *Unlikely History: The Changing German-Jewish Symbiosis, 1945–2000*. Eds. Leslie Morris and Jack Zipes. Houndmills: Palgrave Macmillan, 2002. 291–306. Print.

Oliver, Max. Rev. of *Her Father's Daughter*, by Alice Pung. *Bookseller + Publisher* August, 2011: 24. Print.

Ommundsen, Wenche. "'This Story Does Not Begin on a Boat': What is Australian about Asian Australian Writing?" *Continuum* 25.4 (2011): 24. Web. 29 Jan. 2016.

On, Thuy. "Filial Love Song." Rev. of *Her Father's Daughter*, by Alice Pung. *Australian Book Review* 24 Sept 2011: 24. Print.

Owen, Jan. "Memories and Dismemories." Rev. of *Her Father's Daughter*, by Alice Pung. *Quadrant* March, 2012: 106–7. Print.

Phung, Malissa. "The Diasporic Inheritance of Postmemory and Immigrant Shame in the Novels of Larissa Lai." *Postcolonial Text* 7.3 (2012): 1–19. Web. 13 Feb. 2016.

Pung, Alice. *Unpolished Gem*. Melbourne: Black Inc., 2006. Print.

Pung, Alice. *Her Father's Daughter*. Melbourne: Black Inc., 2011. Print.

Pung, Alice. Interview by Anita Barraud. *The Book Show*, ABC Radio National. 13 Sept. 2011. Print. Transcript.

Pung, Alice. *Laurinda*. Melbourne: Black Inc., 2014. Print.

Pung, Alice. "No Subject Line." *Message to Anne Brewster*. 4 Feb. 2016. E-mail.

Stocks, Claire. "Trauma Theory and the Singular Self: Rethinking Extreme Experiences in the Light of Cross Cultural Identity." *Textual Practice* 21.1 (2007): 71–92. Taylor & Francis Online. Web. 12 Feb. 2016.

Sullivan, Jane. "Memories of Relative Unease." *The Age*. Fairfax Media, 11 Aug. 2011. Web. 30 May 2016.

Themonthlyvideo. "Her Father's Daughter. Alice Pung in Conversation." Podcast. *TheMonthly.com.au*. YouTube, 2 May 2013. Web. Web. 30 Jan. 2016.

Walker, Brenda. "Noted: Her Father's Daughter." Rev. of *Her Father's Daughter*, by Alice Pung. *The Monthly* Sept, 2011: 64. Web. 1 Jun. 2016.

Whitlock, Gillian. *Postcolonial Life Narrative: Testimonial Transactions*. Oxford: Oxford University Press, 2015. Print.

Witnessing Moral Compromise: 'Privilege', Judgement and Holocaust Testimony

Adam Brown

ABSTRACT
This article examines the complex intersection(s) of representation and moral judgement in the context of Holocaust testimony. The untold traumas and ethical dilemmas confronting Jews *in extremis* remain a difficult subject to engage with in any medium, and the recollections of those who themselves held 'privileged' positions in the ghettos and concentration camps pose specific and important challenges. Drawing on the influential writing of Auschwitz survivor Primo Levi and taking Benjamin Jacobs's memoir *The Dentist of Auschwitz* (2001) as a central case study, I position judgement as a key feature that needs to be explicitly and self-consciously exposed within an ethical framework of reading and understanding.

> It was every man for himself. Who else would be for him? [*sic*] ... Although I still existed in a breathing, seeing, hearing, functioning body, no sane person would have bet a penny on its getting through the next day. (Frister 287–88)

The above passage from Auschwitz survivor Roman Frister's memoir *The Cap: The Price of a Life* is representative of sentiments in countless other Holocaust testimonies. The rhetorical separation of person and body signals the state of untold numbers pushed to the limits of human existence – victims who, at the threshold of death, were disparagingly referred as *Muselmänner*[1] by fellow prisoners. Not only does Frister gesture to the indescribable physical agony inflicted upon prisoners who were always intended to die, but also to the ethical consequences of being incarcerated in what many have viewed as an environment in which conventional standards of morality no longer existed. This Hobbesian scenario, where victim was turned against victim, was the direct product of a complex prisoner hierarchy constructed and enforced by the Nazi persecutors. Jews found themselves at the bottom of the hierarchy, though even here there were layers of 'privilege' from which even they could benefit in life-prolonging ways.

Even though they pervade survivor stories at every turn, to a large degree 'privileged' prisoners have remained an understudied phenomenon. When I was told several years ago by a guide and daughter of survivors at the Sydney Jewish Museum that the subject was 'too difficult' to talk about with visitors, this reinforced the impression that the issue of 'privileged' Jews has remained taboo in many contexts. A growing interest in these figures by filmmakers and recent scholarship on the representation of 'privileged' Jews in film and other media has highlighted a need to delve deeper into the ethical dilemmas confronting Holocaust victims and how they portrayed – whether by themselves or others.[2] This article investigates the ways in which trauma, judgement and compromise intersect in life writing by Jews who held 'privileged' positions in the camps. Drawing on Primo Levi's paradigmatic concept of the 'grey zone', I focus in particular on Benjamin Jacobs's memoir *The Dentist of Auschwitz* (2001), examining the ways in which he represents prisoner-functionaries in the concentration camps along with his own 'privileged' position.

Survivor narratives, 'privileged' Jews and the 'grey zone'

> Every victim is to be mourned, and every survivor is to be helped and pitied, but not all their acts should be set forth as examples. (Levi 9)

The use of the phrase 'life writing' might seem in some ways to be oxymoronic in the context of Holocaust testimonies, which have attracted immense scholarly interest over the years (see, for example, Bigsby; Bernard-Donals and Glejzer; Wieviorka). Bearing witness to systematic, industrialised death on an unprecedented scale, survivor memoirs and video testimonies invariably move quickly from a brief account of pre-war life to descriptions of torture and torment in ghettos and/or concentration camps. Accounts of liberation frequently lack the elation and catharsis afforded by a number of Hollywood films. Many survivors' realisation of the loss of most or all of their family dawned on them only after their extreme dehumanisation began to abate. Oftentimes, amidst ongoing persecution and suffering, months would pass before survivors reached home (if they still had one), and even then most found themselves repressing the past in order to cope in an indifferent world. When physical recovery took hold, many continued to be afflicted by 'survivor guilt' in the general sense of living while so many others did not, if not also feeling shame more specifically in relation to their own behaviour *in extremis*. This last factor is particularly pertinent to Primo Levi's highly influential reflections on the 'grey zone'.

An assimilated Italian Jew who trained as a chemist and briefly joined the resistance against the Nazi occupation of Italy in 1943, Levi was deported to Auschwitz in February of the following year. He has become one of the most well-known survivors, not least due to his highly literary accounts of his wartime experiences in the memoirs *If This is a Man* and *The Truce* (first published in 1947 and 1963 respectively). Levi became a prolific writer of short stories, essays, poetry and a novel after liberation, and has since become the subject of several biographies himself (Angier; Anissimov; Thomson) – not to mention the mass of Holocaust scholarship that has been informed by his work. Even when his writing did not overtly concern the Holocaust, the nature of what it is to be human was never far from Levi's mind. While many scholars point to the deeply humanist foundations of Levi's thought, others have argued that the Holocaust had completely 'sabotaged the ethical vision that [Levi] cherished as a human being' (Langer, 'Legacy' 198). This tension would run alongside his growing preoccupation with the issue of 'privileged' Jews until the end of his life.

Levi's concept of the 'grey zone' developed gradually over the years, moving from a general concern about 'moral compromise' to a conceptual framework underpinning his desperate attempt to understand the camps and what they revealed about human nature. Explicated most clearly in an essay in his final memoir *The Drowned and the Saved* (1986), published shortly before he took his own life, Levi wrote that the 'grey zone' comprises 'the space that separates (and not only in Nazi Lagers) the victims from the persecutors' (25). He took care to emphasise that blurring boundaries in this way should not undermine them: while perpetrators must be held accountable for their actions, he argued that judgement should be suspended in relation to the behaviour of their victims. I've argued elsewhere that Levi could not hold himself to this standard, as the language, metaphors and allusions he uses to portray the actions of various groups and individuals throughout his oeuvre invariably pass judgement on them – even if it is more nuanced than most (Brown, *Judging*). In large part a metaphor for moral ambiguity, the 'grey zone' also at times seems to be conceptualised as a spectrum along which individuals and groups can be judged. While Levi's reflections were often filled with paradox and contradiction, he seemed to hold that the behaviour of 'privileged' Jews should be neither condemned nor (as highlighted in the epigraph to this section) glorified.

Categories of 'privileged' Jews include, for example, the *Judenräte* (Jewish councils) and *Ordnungsdienst* (Jewish police) of the ghettos and *Blockälteste* (block elders) of barracks and *Kapos* (chiefs) of labour squads in the camps, among many others in the complex prisoner hierarchy. Writing of the *Sonderkommandos* (special squads) forced to work in the death camp crematoria, Levi stressed that 'here one hesitates to speak of privilege' (*The Drowned* 34), though fundamental questions of agency and responsibility apply to all. Likewise, all such positions afforded certain benefits, ranging from extra rations, to protection from manual labour in the elements, to relative 'safety' from being deported to camps or 'selected' for the gas chambers. These advantages can only be described as *life-prolonging* rather than *life-saving* given that, in the case of Jewish victims, all were intended to die under the Nazis' genocidal policy. Reflecting the extreme situations that arose in this environment, Jewish responses to their unprecedented persecution (and not only those who held 'privileged' positions) have been conceptualised by Lawrence L. Langer as being governed by 'choiceless choices' – 'crucial decisions [that] did not reflect options between life and death, but between one form of abnormal response and another, both imposed by a situation that was in no way of the victim's own choosing' (*Versions* 72). The complexities of navigating the intersection of coercion and compromise under Nazi oppression would seem to lend much credence to Levi's stance on being 'paralysed' by an '*impotentia judicandi*' (*The Drowned* 43, 49). Nonetheless, while Levi argued that judgement in relation to 'privileged' Jews should be 'suspended' (43), evaluations permeate Holocaust representations across media of all kinds, and survivor testimony is no exception.

Most Jews who held 'privileged' positions did not survive the Holocaust, and many of those who were liberated have not told their story. The task of accounting for the actions of the 'privileged' has often fallen to those who witnessed them from the perspective of 'non-privileged' prisoners. This has frequently involved the portrayal of a prisoner's so-called 'collaboration' by an author who feels that they personally or that their fellow inmates were harmed as a direct consequence of the subject's decisions and behaviour. Negative judgements are therefore (and understandably) common, though some recent attention to survivor literature has pointed to a diverse array of judgements evoked (Brown, "Traumatic Memory"). While few in number, testimonies by former 'privileged' Jews have increasingly informed Holocaust scholarship and memory culture, such as John K. Roth's philosophical reflections on Calel Perechodnik's diary of his time in the Jewish ghetto police and the widespread release of memoirs like *Sonderkommando* member Filip Müller's *Auschwitz Inferno* (1979), Auschwitz orchestra inmate Fania Fénelon's *The Musicians of Auschwitz* (1999), and the posthumous diary of the Jewish elder of the Warsaw Ghetto Adam Czerniakow (1979). As Susan Pentlin argues in her essay 'Holocaust Victims of Privilege', it is crucial that one listens to the 'voices from the grey zone' by exploring the often taboo issues of 'position and privilege' (39, 26).

Writing of the fragmented manuscripts buried by men who, with few exceptions, perished as members of the *Sonderkommando* in Auschwitz-Birkenau, Levi wrote that 'from men who have known such extreme destitution one cannot expect a deposition in the juridical sense of the term, but something that is at once a lament, a curse, an expiation, and an attempt to justify and rehabilitate themselves' (*The Drowned* 36). This sentiment holds to varying degrees for testimonies of 'privileged' prisoners more broadly, as can be seen in the lesser-known experiences of prisoner-doctors – a group of victims often trapped in an ambiguous position between duty of care and complicity. Exemplifying this is German prisoner-doctor Ella Lingens-Reiner's reflection on the dilemma she faced in trying to use her 'privileged' position in Auschwitz to help others amidst the Nazis' obsession with numbers:

> If I rescued one woman, I pushed another to her doom, another who wanted to live and had an equal right to live ... Was there any sense in trying to behave decently? (Lingens-Reiner 82)

Sufficient attention has yet to be given to prisoner-doctors as a category of 'privileged' prisoners, though several memoirs by Jewish survivors lend themselves to an exploration of the ethical dilemmas they and others around them faced.

Personal narratives of former 'privileged' Jews provide a valuable prism through which to analyse the ethics of Holocaust life writing – both the ethics of writing these narratives and the ethics of reading them. Indeed, such examples may be seen to represent *limit cases* of life writing, where the paucity of language – to describe human behaviour and experience, much less provide a means to evaluate it – is revealed to be particularly acute. With this in mind, I take what John K. Roth identifies as a 'metaethical' approach: a reflection on judgements that have already been made, which 'seek[s] to understand more fully how those judgments work as well as what limits they face and problems they entail' (60). I would argue that this approach is not purely – or ideally – an academic one, but exemplifies something that should be part of any reader's engagement with Holocaust narratives (and not only personal ones as with survivor testimony, or literary narratives alone). Indeed, this problem can be situated within the field of life writing more broadly, reflecting G. Thomas Couser's concerns regarding the ethical obligations of an author when representing 'vulnerable subjects', who 'are unable to represent themselves in writing or to offer meaningful consent to their representation by someone else' (xii). A close reading of former prisoner-doctor Benjamin Jacobs's memoir *The Dentist of Auschwitz* highlights the importance of exposing and exploring the presence of moral judgement(s) in Holocaust life writing.

'The difference between life and death': the dentist of Auschwitz

The positions of prisoners working in the so-called 'hospitals' of Auschwitz often afforded opportunities to resist the Nazis (Strzelecka 391), though to appropriate Levi's metaphor further, shades of grey can be found in many ambiguous figures trapped in these impossible scenarios. A comprehensive picture of the diverse roles played by prisoner-doctors in the Nazi-controlled camps is beyond the scope of this article, but a few brief examples serve to map out the sensitive ethical terrain a reader must navigate when engaging with their narratives. Two prisoner-doctors whose stories have become recognised enough to be figured in filmic treatments are Gisela Perl and Miklós Nyiszli.[3] Both Hungarian Jews, Perl and Nyiszli were incarcerated in Auschwitz-Birkenau where they found themselves in the complicated positions of providing medical treatment to both inmates and SS guards. Coming into contact with the notorious Josef Mengele, both were forced to assist in different ways with the Nazi doctor's pseudo-scientific 'research'. Finding a significant means of resistance, Perl describes in her memoir the trauma she experienced in secretly performing a large number of abortions to save women from being sent to the gas chambers. At one point, she ambiguously describes the 'beautiful specimen' of an eight-week-old foetus she provided to Mengele (119–20). By comparison, Nyiszli was given the task of conducting dissections of the bodies of certain prisoners Mengele found 'of interest'. Nyiszli's position afforded him a variety of benefits, including at one point being able to get his wife and daughter transferred to a less severe camp setting. Ordered by Mengele to ensure he did a 'good job', Nyiszli writes in his memoir that he 'planned to carry out his orders to the best of my ability' (34). His apparent diligence gave one survivor cause to describe Nyiszli and Mengele as 'very comfortable together' (Lifton 370).

Benjamin Jacobs (born Berek Jakubowicz and renamed after the war) became a prisoner-doctor of a very different kind in Auschwitz, but his experiences at and long before his incarceration there bear striking resemblances to those of many other 'privileged' Jews. Like many other survivors, Jacobs only recorded his story for posterity late in life, having decided to write his memoir after participating in a 'fact-finding' trip to Europe on behalf of the United States in July 1985 and after surviving a cancerous tumour. The perspective of someone looking back from afar differs greatly from, for instance, Nyiszli's recollections set down in the immediate post-war period (he died in 1956). The acknowledgements in Jacobs's preface underline his commitment to researching the events and settings he lived through, and his reliance on documentation at certain times is an important feature of his memoir to which I will return. Like Levi's various testimonial writings, Jacobs's memoir is often as much about *understanding* what happened as *retelling*, particularly in relation to the human behaviour(s) he witnessed – and judgement is a key feature of this representational process.

Jacobs benefited from one year of formal dental training in several settings as he was moved from camp to camp by his persecutors, with this arguably helping to facilitate (in part) his survival until the end of the war. Based on his work in previous camps, Jacobs was given the opportunity (without seeking it out) to take up the position of dentist at Fürstengrube, the sub-camp of Auschwitz III (Buna-Monowitz) in which he was incarcerated. While dentistry was far from his only activity in the camps, hygiene and dietary conditions meant there was no shortage of patients requiring his services. Jacobs most frequently removed teeth, often with no anaesthetic, and treated diseased gums with iodine. The place of 'privilege' in Jacobs's story is constantly foregrounded; indeed, he opens his first chapter by referring directly to the role dentistry would play: 'Little did I know then that those tools would save my life' (1). On the other hand, Jacobs begins his memoir with more emphasis on others who held 'privileged' positions, and as he moves forward many implicit contrasts can be found – or at the very least made – between Jacobs's reflections on his own behaviour and that of other 'privileged' prisoners around him. The narrative is interlaced with accounts – and judgements – of various figures who sometimes occupy passages in a similar way to Levi's own use of literary vignettes in his testimonies.

Kurt Goldberg, the *Lagerältester* (camp leader) in Gutenbrunn concentration camp, is one such example. A child of a 'mixed marriage' who had joined the 'Hitler Youth' before Nazi race laws had seen him reclassified as Jewish, Goldberg remains for Jacobs 'the most enigmatic' of the 'many diabolical characters anointed in that era' (78). Describing him as particularly (and inexplicably) 'cold and callous' toward his fellow Jewish prisoners (77), Jacobs ponders Goldberg's motivation:

> [H]e was convinced that he deserved better, and he took out the anger and frustration of his misfortune by intimidating his fellow Jews. His boldness and his command of the German language made him a perfect tool for the malevolent Nazi system. (78)

Curiously, no reference is made in Jacobs's account to the possibility or likelihood that personal survival also played a role; however, the representation of Goldberg becomes more nuanced when Jacobs goes on to describe his personal interactions with Goldberg. Having become the camp dentist after offering his services to Goldberg, Jacobs describes the development of 'a smoother relationship, and he even went out of his way to help me' (80). The vulnerability of as seemingly powerful a figure as Goldberg becomes clear when he loses his authority to another Jewish prisoner, Richard Grimm, who aspires to replace him and succeeds by courting the SS (85–86). Revealing his own resourcefulness, Jacobs would also develop a 'good relationship' with the latter, while Goldberg 'sank into isolation' (97, 87). Later in his memoir, Jacobs recalls being visited by Goldberg in Auschwitz. He writes that the former *Lagerältester*, 'a scant image of the man we had once feared', had wanted to 'unburden his soul' to Jacobs: Goldberg expressed profound regret over abandoning his cultural heritage and spoke in perfect Yiddish to him (158). Through this passage in particular, Jacobs positions himself as a witness in more than one way – not only in terms of being one who recounts events and the people caught up in them, but in this case as the (presumably sole) confessor to a complex human being who died before the war ended and hence never had opportunity to tell his own story (if indeed he would have chosen to).

Jacobs's memoir points to a number of individuals with varying backgrounds and behaviours – oftentimes, the description is short, but some form of moral evaluation is invariably expressed. This can be seen not least of all in Jacobs's portrayal of the *Kapos* (chiefs) of labour squads in Auschwitz. He writes they were 'arrogant and harshly indifferent' despite the fact 'they faced the same life that we did' (126). As in the case of Goldberg and Grimm, Jacobs generally distinguishes between individual prisoner-functionaries, forming a complex network of umbrella judgements of different categories of 'privileged' Jews with individual exceptions noted throughout. Immediately after identifying one prisoner as 'the friendliest and the most decent' of all the assistants to a *Kapo* he encountered,

Jacobs writes: 'The Kapo, however, was different. When he began to speak, he demonstrated how, in Auschwitz, men become more aggressive than animals. He looked well-nourished. He laid down the rules' (127). The rhetorical shift between the general and particular in survivor testimony can be immensely loaded. The implicit distinction made toward the end of Jacobs's memoir of '540 inmates plus the Kapos' is particularly telling (177). Jacobs's fluctuation between stern judgement and acknowledgement of the mortal dangers that 'privileged' prisoners were constantly exposed to reveals the difficulty in evoking their experiences, much less evaluating them. To complicate matters further, Jacobs also seeks to hold in balance his own position(s) in the camps and the benefits these exposed him to.

Upon being transferred from the Ghetto in Dobra to the Steineck concentration camp, Jacobs writes of the 'deeply dehumanizing effect' of the traumatic initiation period in the camp for all prisoners, including himself. While he unquestionably suffered in many ways, Jacobs found himself 'fortunate' enough to almost immediately be granted a position that relegated him to a place in the prisoner hierarchy slightly above most others. Along with (and as a result of) the discovery of his dentistry ability, Jacobs had the low but not entirely insignificant role of *Kolonnenführer* ('group leader') conferred on him. He writes little of what this role specifically involved, though the benefits he received as a result of people he interacted with soon become clear, including extra rations and work responsibilities that were considerably easier to cope with than the manual labour the majority endured. Even from this early point in his recollections, Jacobs expresses an awareness of the ethical dilemmas he faced. Describing an initial reluctance to accept the extra food the camp's cook offered him, Jacobs writes: 'With my fellow inmates eating from the kettles, how could I, in their presence, have different food? … In time, however, hunger won out, and I occasionally accepted her leftovers' (47). Further, Jacobs secured a role carrying water (which even provided adequate 'freedom' of movement to engage in a love affair with a Polish woman working nearby), and an even more significant administrative position that 'brought me some influence' while other prisoners broke rocks (48). Jacobs was also able to use his contacts to secure his weakening father a job as potato peeler. At the same time, he recounts being actively involved in efforts to help those around him, undertaking activities that would by many definitions be defined as resistance. These included a key role in a dangerous scheme to procure bread from a bakery for the camp's workforce, which saw him whipped almost to death and officially demoted from the role of *Kolonnenführer* – although the functionaries who held him in favour ensured he kept his administrative position (63). Such activities helped himself and those closest to him – there was seldom the possibility of aiding everyone.

Jacobs's 'privileged' perspective (in both senses of the word) raises important questions for those who engage with his testimony. On the one hand, he recollects events involving individuals who for the most part did not survive and cannot speak for themselves. On the other hand, Jacobs's 'privileged' position and perspective to a degree separates him from other victims' experiences. He displays an acute awareness of this at certain times, particularly when writing of the growing sense of being looked upon differently by other prisoners: 'In time I realized that my having been placed in a position that gave me an advantage carried with it rejection by the other inmates' (48). Jacobs's dentistry even afforded him the ability to travel between camps (with a guard) by offering his dental services, allowing him to confirm the fate of his mother and sister in the gas vans of Chelmno. More than once

Jacobs expresses a mixture of embarrassment and shame over the distance between himself and the non-privileged prisoners around him: 'The whole group was looking at me, wondering how I could ride around without the mark of a Jew' (96). Comparing himself explicitly with his fellow inmates, he later writes 'I was the lucky one ... being the dentist made my day-to-day life more bearable' (103). As a result, Jacobs's testimony often fluctuates between observation and direct experience, depending on the distance his 'privileged' positions keep him from the camp's everyday torments. He writes empathetically to be sure, but there is a significant difference between reading his description of the 'heartbreaking' sight of 'some of our people standing ankle-deep in mud, lifting fourteen-kilo shovelfuls of dirt' from Jacobs's viewpoint and reading the words of those who themselves suffered from (to necessarily draw on Jacobs's words again) 'calloused and cracked' hands and 'open wounds' (68). Clearly, no matter how crucial Jacobs's testament to past horrors is, the difference between witnessing first-hand and experiencing first-hand holds some significance here.

Another form of distance – in this case, temporal – has a strong impact on the construction of Jacobs's memoir, as he clearly incorporates details throughout that he has accumulated from sources other than first-hand memory alone. Common to survivor testimonies articulated long after the events recalled, a simple example of this is Jacobs's comment early in his story that 'We were resigned to Churchill's prediction: "This will be a long war"' (69). Of course, the source material that may have impacted (knowingly or not) on a survivor's account is not always clear, thus while Jacobs does not refer to the concepts of the 'grey zone' or 'choiceless choices' explicitly, there is no way of ruling out that other assessments of the ethical dilemmas confronting 'privileged' Jews by other survivors, historians or even films have informed his narrative. Indeed, Jacobs does show an awareness of the wider and massively researched issue of 'Holocaust representation' when he recalls thinking, on his arrival in Auschwitz, that 'no epic drama could duplicate the sight that was before me. No one would be able to find such emaciated bodies to re-create the scene' (122). Signalling a rare practice in Holocaust testimony, Jacobs goes even further than using brief, oblique references that are unlikely to have stemmed from personal knowledge during the war, drawing explicitly on several other published sources at length. He quotes a passage from another survivor's testimony for half a page to fill in his lack of direct insight regarding what happened to prisoners left in Fürstengrube after he departed on a 'death march' (165). Jacobs also relies on the writing of others, including Holocaust historian Martin Gilbert, to fill in other details of his later experiences (192–93). This practice reveals an intention to provide as comprehensive a picture as possible, which may help explain at least in part his preoccupation with examining the wider spectrum of prisoner experiences and behaviours beyond his own. Arguably, the benefits of Jacobs's 'privileged' position provided him with an opportunity to observe this spectrum, just as he (re)constructs it within his memoir. Yet Jacobs's own tenuous position sees his memoir negotiate shifting levels of vulnerability and status within the camp hierarchy.

Upon being deported to Auschwitz, which included days of starvation and other torments in a cattle wagon, Jacobs was stripped of the 'privileges' he once held – to the point that his dental instruments were literally thrown on the train platform by an SS guard. While an extremely close call during the initial camp 'selection' leads Jacobs to declare that 'survival, all else aside, was primarily luck' (121), he freely admits the many benefits of once again becoming a camp dentist in Fürstengrube:

> I was given an elite camp suit, a sweater, and a pair of real leather shoes, which distinguished me from the Kommando inmates. I also continued to receive kitchen privileges. I stopped being the dumb inmate and no longer needed to fear the Kapos or the foremen. (142)

The camp's dental station was situated adjacent the 'penal room', where Jacobs could hear the suffering of those tortured within – providing a valuable metaphor for the barrier that 'privilege' constructed between him and the more vulnerable (143). As before, this barrier was not made of concrete and served only as transitory protection. For four months, Jacobs was constantly tormented and tortured by *Unterscharführer* Günther Hinze, a particularly anti-semitic SS guard who forced him to complete physical stress exercises on a regular basis (144–45). Ironically, Jacobs's account of Hinze declaring 'You have it too good here' gestures to the possibility that it was the victim's 'privileged' position that served as an excuse for the torture. Jacobs is openly self-conscious of the benefits of the 'privileged' position(s) he held, describing his dental tools at one point as having 'magical powers' (175) and the Fürstengrube dental station as 'my security and my torture chamber' for almost 18 months (164). The visibility and apparent respect that Jacobs's position gave him proved valuable on multiple occasions after leaving Auschwitz, with Jacobs even given vodka to keep him walking on the 'death march' westwards while others who could no longer continue were shot (167–68).

As with Perl and Nyiszli's interactions with Mengele, one of the more provocative aspects of the experiences of 'privileged' Jews proved (and still proves) to be their relations with the SS. Jacobs's position saw him treat the SS officers who ran the camp and brought him into contact with *Hauptscharführer* Otto Moll, a notorious German officer deeply involved in the killing process. Recalling that Moll provided him with extra rations and allowed him to forego forced labour for his hands to recover in order to better serve as dentist, Jacobs writes that 'I would never forget the human, almost tender way in which he spoke' (142). He even notes that Dr Schatz, a disillusioned SS officer, 'acted as if we were equal', which undoubtedly did not apply to all prisoners in the camp (150). Jacobs recounts very nervously treating Moll for a cavity at gunpoint, a task that resulted in him securing better jobs for his father and brother, who left their labour in a coal mine to work as barracks and 'hospital' orderlies (151). Furthermore, Jacobs was able to exploit his contacts to have Moll transfer his aforementioned tormentor Hinze elsewhere, revealing the remarkable influence he held in contrast to most other prisoners. Yet with influence and advantage came mixed feelings: on witnessing a frenzied mass shooting by Moll one night from the 'safety' of his dentistry workplace, Jacobs notes 'I could not bear looking out any longer … I felt a deep shame that my circumstances had separated me from those whose fate I shared' (153).[4] While Jacobs openly alludes to the ethical dilemmas he confronted in earlier camps, he separates himself from other 'privileged' prisoners in Auschwitz in both subtle and, in the following passage, explicit ways:

> Even the worst, most menial job in camp could make the difference between life and death. Because of that, many prisoners were anxious to take any camp job, even if it meant helping the Nazis. Fortunately the camp dentist did not have the same dilemma. (142)

While 'helping the Nazis' could be defined in many ways, certain recollections throughout Jacobs's account of his time in Auschwitz might not reflect the above sentiment in the eyes of some. On one occasion, Moll used Jacobs's dental station as 'a perfect example of the many kinds of health care' provided by the Nazis when deluding Red Cross representatives who visited the camp (155). In fact, the SS exploited Jacobs's talents for not only propaganda but also for economic value.

In what would prove one of the more traumatic aspects of Jacobs's incarceration, he was required to pull gold fillings from dead prisoners. This activity gives rise to Jacobs's most sustained rumination on the 'choiceless choices' he faced:

> I felt revulsion. I did not think that anyone could stoop that low. Killing people was horrible enough, but tearing out teeth of the dead moved me to disgust. I did not think that I could do it. But it was inevitable. I had no choice. … It was by far the hardest thing I had to do in any camp. I often asked myself what would have befallen me if I had not complied with this order. I have never stopped wrestling with that question. When I approached the corpse room for the first time I tried to rationalize that what I was about to do was meaningless to the dead. But it never was to me. (147)

The gold was to be used to make bridges and caps for the SS officers, though some was also allocated to make jewellery for the guards' own personal enrichment. In turn, Jacobs was treated with 'modified' behaviour by the guards, who provided him with 'bread, sausages, and cigarettes, while cautioning me not to tell anyone' (148). He immediately notes that these 'offerings went a long way toward helping me, my father, my brother, and a number of other inmates' (148). Jacobs provides a detailed description of the gruesome sight that confronted him on each visit to the morgue and the shame that resulted: 'My father and my brother also knew what I was doing, though I never told them' (148). Notably, Jacobs secretly took a bag of dental gold with him after Auschwitz was abandoned, though lost it before liberation.

Speaking for the dead and the ethics of reading

Dominick LaCapra emphasises in his chapter 'Holocaust Testimonies: Attending to the Victim's Voice' that a key issue to reflect on in relation to testimonies is how to 'come to terms with affect in those who have been victimised and traumatised by their experiences, a problem that involves the tense relation between procedures of objective reconstruction of the past and empathic response, especially in the case of victims and survivors' (85). This tension is particularly acute when confronted with the behaviour of 'privileged' prisoners and those who represent them. Readers must respect the trauma 'privileged' Jews themselves endured, but self-consciously acknowledge the inevitability of moral judgement – both their own and those who testify. Robert Rozett writes that Jewish witnesses' firsthand accounts cannot clarify the Nazis' motives or plans, but 'can only teach us about the effect of the horror on the individual victims and the experiences of the victims facing the horror' (100–101). This acknowledgement of perspective does not detract from the crucial importance of survivor accounts; however, an awareness of the subjective nature of diaries, memoirs and oral testimonies raises important ethical questions when considering how survivors represent their own or others' experiences.

Toward the end of his pivotal essay on the 'grey zone', Levi raises a crucial point regarding the controversial Jewish leader Chaim Rumkowski of the Lodz Ghetto that cannot easily be set aside. Like most 'privileged' Jews, Levi argues, the only words about Rumkowski that we lack and can never obtain are his own. Indeed, Levi writes that only Rumkowski could clarify his situation 'if he could speak before us, even lying, as perhaps he always lied, to himself also; he would in any case help us understand him, as every defendant helps his judge' (*The Drowned* 50). In Jacobs's case we do have his story in his words, although in the expansive telling of his story, these words also stand in place of the words of many others just as much as they testify to his own behaviour. Jacobs actually comments on the subject of Jewish leaders himself in the opening pages of his memoir, offering a sweeping judgement of the Jewish Council and ghetto police in Nazi-occupied Dobra as men of 'little conscience' who 'wielded indiscriminate power over us' (2). Shortly afterward, Jacobs does acknowledge that Jewish leaders faced terrible ethical dilemmas and ultimately the same end:

> As difficult as it may have been for them to make the choices that no human should ever have to make, the members of the Judenrat primarily sheltered themselves, their families, and their friends from the privation and discrimination the rest of us had to face. Little did they know that having sent their people to their deaths, they would in the end fall victim to the same fate. (2)

There is some irony in the writer's particularly strong judgement that informs this passage, given that Jacobs's own 'privileged' position(s) benefited himself, his father, brother and others close to him at many turns. This point is not to condemn Jacobs, of course, but to highlight the importance of incorporating a metaethical element into the reading of survivor testimonies, which self-consciously reflects on the passing of judgements in all their forms. In the context of Holocaust life writing in particular, this applies to similar limit cases where difficult issues of 'privilege' and 'compromise' are confronted.

Through the narrative strategy of representing others' experiences and behaviours as much as his own, Jacobs provides a clear impression that he seeks to reconstruct as comprehensive a picture as possible. He modestly writes in a footnote to his 'Postscript' that 'I am the least important person in this book. It is the memories of the events that overtook us that must be remembered' (215). Along with their personal, humanising perspective and 'relatability', the inextricable connection between the individual and the larger settings, events, and stories they find themselves caught up in, suffering, and enduring is what gives survivor accounts the power they have. The limits of subjectivity must be acknowledged, to be sure, and the problem of judging victim experiences *in extremis* remains a vexed one. The issue of 'privileged' Jews poses acute challenges for the reader at many turns throughout Jacobs's memoir, which simultaneously describes and evaluates; at once judges and points to unresolvable ethical dilemmas that would seem to call into question the appropriateness of judgement itself.

Notes

1. *Muselmann*, the German term for Muslims, was used to categorise prisoners in the concentration camps who appeared to be overwhelmed by starvation and despair, and seemed to be beyond all help and hope of survival.
2. Recent films that portray 'privileged' Jews include László Nemes' *Son of Saul* (2015) and Stefan Ruzowitzky's *The Counterfeiters* (2009) – both of which won Academy Awards for Best Foreign Language Film – which followed earlier productions such as Audrius Juzenas' *Ghetto* (2006), Joseph Sargent's *Out of the Ashes* (2003), and Tim Blake Nelson's *The Grey Zone* (2001).
3. Perl was the protagonist of *Out of the Ashes*, whereas the figure of Nyiszli is more indirectly represented in both *The Grey Zone* and *Son of Saul*. Both survivors are discussed at length elsewhere (Brown *Judging*, "'No One Will Ever Know'").

4. Following this massacre of 19 prisoners, Moll allowed Jacobs to sit with his dying father and then say Kaddish over his body, exemplifying that human complexity did not rest with the victims alone (161–62).

Disclosure statement

No potential conflict of interest was reported by the author.

References

Angier, Carole. *The Double Bond: Primo Levi, A Biography*. London: Penguin, 2003. Print.
Anissimov, Myriam. *Primo Levi: Tragedy of an Optimist*. Trans. Steve Cox. London: Aurum, 1998. Print.
Bernard-Donals, Michael, and Richard Glejzer. *Between Witness and Testimony: The Holocaust and the Limits of Representation*. Albany: State University of New York Press, 2001. Print.
Bigsby, Christopher. *Remembering and Imagining: The Chain of Memory*. Cambridge: Cambridge UP, 2006. Print.
Brown, Adam. *Judging 'Privileged' Jews: Holocaust Ethics, Representation, and the 'Grey Zone'*. New York: Berghahn, 2013.
Brown, Adam. "'No One Will Ever Know … ': The Holocaust, 'Privileged' Jews, and the 'Grey Zone'." *History Australia* 8.3 (2011): 95–116.
Brown, Adam. "Traumatic Memory and Holocaust Testimony: Passing Judgement in Representations of Chaim Rumkowski." *Colloquy: Text, Theory, Critique* 15 (2008): 128–144.
Couser, G. Thomas. *Vulnerable Subjects: Ethics and Life Writing*. Ithaca: Cornell University Press, 2004.
Czerniakow, Adam. *The Warsaw Diary of Adam Czerniakow: Prelude to Doom*. Eds. Raul Hilberg, Stanislaw Staron, and Josef Kermisz. New York: Stein and Day, 1979. Print.
Fénelon, Fania, *The Musicians of Auschwitz*. Trans. Judith Landry. London: Sphere, (1977) 1999. Print.
Frister, Roman. *The Cap: The Price of a Life*. Trans. Hillel Halkin. New York: Grove, (1993) 1999. Print.
Jacobs, Benjamin. *The Dentist of Auschwitz: A Memoir*. Lexington: The UP of Kentucky, 2001. Print.
LaCapra, Dominick. *Writing History, Writing Trauma*. Baltimore: The Johns Hopkins UP, 2001. Print.
Langer, Lawrence L. "Legacy in Gray." *Memory and Mastery: Primo Levi as Writer and Witness*. Ed. Roberta S. Kremer. Albany: State University of New York Press, 2001. 197–216. Print.
Langer, Lawrence L. *Versions of Survival: The Holocaust and the Human Spirit*. Albany: State University of New York Press, 1982. Print.
Levi, Primo. *The Drowned and the Saved*. Trans. Raymond Rosenthal. London: Michael Joseph, (1986) 1988. Print.

Levi, Primo. *If This is a Man; and, the Truce*. Trans. Stuart Woolf. London: Abacus, (1979) 1995. Print.

Lifton, Robert Jay. *The Nazi Doctors: Medical Killing and the Psychology of Genocide*. Basic Books, (1986) 2000. Print.

Lingens-Reiner, Ella. *Prisoners of Fear*. London: Victor Gollancz, 1948. Print.

Müller, Filip. *Auschwitz Inferno: The Testimony of a Sonderkommando*. Trans. Susanne Flatauer. London: Routledge and Kegan Paul, 1979. Print.

Nyiszli, Miklos. *Auschwitz: A Doctor's Eyewitness Account*. Trans. Tibere Kremer and Richard Seaver. New York: Arcade, (1960) 1993. Print.

Pentlin, Susan L. "Holocaust Victims of Privilege." *Problems Unique to the Holocaust*. Ed. Harry James Cargas. Lexington: UP of Kentucky, 1999. 25–42. Print.

Perl, Gisella. *I Was a Doctor in Auschwitz*. Salem: Ayer, (1948) 1992. Print.

Roth, John K. *Ethics During and After the Holocaust: In the Shadow of Birkenau*. Houndmills: Palgrave Macmillan, 2005. Print.

Rozett, Robert. *Approaching the Holocaust: Texts and Contexts*. London: Vallentine Mitchell, 2005. Print.

Steinberg, Paul. *Speak You Also: A Survivor's Reckoning*. Trans. Linda Coverdale and Bill Ford. New York: Henry Holt and Company, (1996) 2000. Print.

Strzelecka, Irena. "Hospitals." *Anatomy of the Auschwitz Death Camp*. Eds. Yisrael Gutman and Michael Berenbaum. Bloomington: Indiana UP, 1998. Print.

Thomson, Ian. *Primo Levi: A Life*. New York: Henry Holt and Company, 2002. Print.

Wieviorka, Annette. *The Era of the Witness*. Trans. Jarod Stark. Ithaca: Cornell UP, 2006. Print.

'A Thing May Happen and be a Total Lie': Artifice and Trauma in Tim O'Brien's Magical Realist Life Writing

Jo Langdon

ABSTRACT
Depicting patently fantastic episodes alongside the traumatic events of the Vietnam War, Tim O'Brien's autobiographical magical-realist writing problematises pictures of the past and challenges conventional generic distinctions. Despite the often impossible and outrageous occurrences that punctuate his otherwise realistic narratives, and indeed the pointed and provocative revisions and contradictions of his self-reflexive texts themselves, O'Brien insists on his work's authenticity. In line with critics such as Eugene L. Arva, O'Brien contends that his writing is true to the 'felt' experience of trauma—which trauma theorists largely characterise according to intensity, non-linearity and confusion. Nonetheless, rather than being pure outpourings of trauma, his works are clearly—and self-consciously—literary artefacts that often resonate with popular culture narratives. Undertaking case studies of the autobiographically informed magical realist novel *Going After Cacciato* and the short-story collection *The Things They Carried*, a text that suggestively functions as both fiction and autobiography, this article attends to the paradoxes of fantasy-filled 'life writing', revealing a convergence between magical realism and testimony when it comes to the representation of trauma.

In their representations of trauma, the genres of life writing and fiction can sometimes intersect in ways that arguably invalidate the traditional division between these categories. Life writing would seem to demand some kind of fundamental truth or reliability, if not a documentary faithfulness to the past. However, understandings of trauma, while contested, largely envisage trauma as inaccessible in ways that pose significant challenges to such expectations. Cathy Caruth's foundational works in the field of trauma studies, for example, contend that as the original experience of trauma and its subsequent telling—its testimony—are marked by silences and gaps, the only way to truly represent trauma is in ways that represent the experience's unattainable elements.

Such a position certainly works to complicate understandings of history, given that history is inevitably a discourse of trauma. 'For history to be a history of trauma,' Caruth writes, 'means that it is referential precisely to the extent that it is not fully perceived as it occurs; or to put it somewhat differently, that history can be grasped only in the very inaccessibility of its occurrence' (Caruth 8). That is, in light of the unspeakable nature of traumatic events and experiences, the referentiality of history and, more specifically, first-hand accounts of traumatic historical events, must be recognised as authentically compromised.

Despite the association of testimonial texts with historical accuracy and representations of 'what happened', Caruth suggests that acknowledging that which resists textual or linguistic representation might in fact make it possible to better understand traumatic experiences and, as a result, to better understand the past. 'Through the notion of trauma,' Caruth contends, 'we can understand that a rethinking of reference is aimed not at eliminating history but at resituating it in our understanding, that is, at precisely permitting *history* to arise where *immediate understanding* may not' (Caruth 11). A more complete picture of the past, then, is one inclusive of the ineffable—one in which elements of blankness, silence or lacuna are given shape or recognition.

In response to these issues, critics such as Eugene L. Arva have proposed a connection between magical realist literature and trauma. According to Arva, magical realism, in which realism and the reality depicted are haunted by the magical, might also be conceived as a kind of trauma literature. Indeed, the mode's political and historical engagement means that magical realist texts invariably depict trauma. This is often done from a postcolonial or feminist perspective, whereby the fantastical episodes underscore the omissions or occlusions of imperialist and patriarchal discourse, those traumatic events overlooked or invalidated by the dominant culture. This makes magical realism a mode of writing that is 'profoundly ironic in its trademark representation of the magical as real' (Takolander and Langdon 4).

In *The Traumatic Imagination: Histories of Violence in Magical Realist Fiction*, a study significant for its focus on this connection between magical realism and trauma narratives, Arva posits that the magical realist text 'gives traumatic events an expression that traditional realism could not ... because magical realist images and traumatized subjects share the same ontological ground, being part of a reality that is constantly escaping witnessing through telling' (6). For Arva, magical realism is uniquely placed when it comes to capturing and conveying traumatic experiences. Of particular interest to this article is that, in its depiction of trauma, magical realism as Arva envisions it might comprise a kind of testimonial writing or life writing.

Tim O'Brien's trauma narratives—examined by Arva as exemplary magical realist trauma texts—are characterised by incongruous flights into fantasy that interrupt his historical representations of the Vietnam War. However, O'Brien himself justifies this fusion of the illusory and the believable, maintaining that 'war is a surrealistic experience' (qtd. in Herzog 22). 'I see myself as a realist in the strictest sense,' he argues. 'That is to say, our daydreams are real; our fantasies are real. They aren't construed as otherwise in any of my books' (qtd. in Herzog 80). The author thus frames the repetitiveness, unreliability and unreal qualities of his narratives as evidence of their authenticity.

Yet, this neutralises the importantly pointed and provocative qualities of his writing. The ironic charge of many of the fantastic elements of his work, along with the repetition of certain tropes and archetypes—references to other texts and popular cultural icons—work to highlight the cultural status of O'Brien's trauma narratives as literary texts, rather than apparent symptoms of trauma. In the words of one of O'Brien's characters in *The Things They Carried*, 'That's how stories work, man' (96). O'Brien's evident awareness of this might place certain 'limits' on his claims pertaining to authenticity. Nonetheless, his writing also constitutes a form of traumatic testimony in ways that pose questions about the limits of life writing.

Theorising trauma and life writing

'Welcome to the contemporary trauma culture,' writes Roger Luckurst in his introduction to *The Trauma Question* (2). This ambivalently inviting statement notes the pervasiveness of traumatic narratives and the accompanying growth of trauma studies. The study of trauma has attracted interdisciplinary attention spanning psychology, psychoanalysis, history and the creative arts. Gillian Whitlock and Kate Douglas refer to the 'burgeoning field of academic writing about trauma' (2). Indeed, the vast array of monographs, scholarly conferences and special journal editions focussing specifically on trauma attest to the prevalent interest in the ways traumatic experiences are witnessed, told, read and understood through different mediums or means of representation—including both nonfiction narratives and creative imaginings. As Leigh Gilmore contends, '[m]emoir is thriving, energized in no small part by a surge in the publication of personal accounts of trauma' (16). According to Gilmore, 'self-authored, first-person narratives of trauma seem to be everywhere: in identity politics, revisions of colonial histories, Holocaust studies, and historiographies focused on violence' (48).

Caruth's emphasis on the difficulty of narrating trauma, as noted, has been particularly influential. For Caruth, patterns of dissociation, blankness and incompleteness, repetition and intensity are so significant to the experience of trauma and its memory that they must be recognised in its telling. Ultimately, these difficulties pertain to language—to verbalising or narrating the traumatic. As Judith Lewis Herman describes it in *Trauma and Recovery: From Domestic Abuse to Political Terror*, 'the story of the traumatic event surfaces not as a verbal narrative but as a symptom' (1). Traumatic events, then, according to many trauma theorists, can only be known incompletely and belatedly, which problematises a clear and complete picture of history. Trauma also makes distinctions between the past and the present difficult, given the ways in which traumatic pasts are understood to repeat—to 'flash' back and haunt survivors in the present.

Notably, recent clinical research and scholarship within the field of literary studies from Richard McNally and Joshua Pederson respectively refute Caruth's vision of trauma as oblique and unspeakable. Indeed, drawing on McNally's work within psychology, Pederson contends that, rather than being 'elusive or absent', traumatic memories are 'potentially more detailed and more powerful than normal ones'—and that therefore 'authors may record trauma with excessive detail and vibrant intensity', such that 'readers looking for representations of trauma may turn not to textual absence but to textual overflow' (339). This view of trauma is certainly germane to magical realism, given the metaphorical intensity and hyperbolic transformations that characterise the mode.

In terms of trauma fiction more broadly, literary depictions of shock and extremity seem typically attuned to the difficulty of trauma's narration. Anne Whitehead suggests that 'fiction itself has been marked or changed by its encounter with trauma', and that '[n]ovelists have frequently found that the impact of trauma can only adequately be represented by mimicking its forms and symptoms, so that temporality and chronology collapse, and narratives are characterised by repetition and indirection' (3). Luckhurst similarly suggests that trauma 'issues a challenge to the capacities of narrative knowledge' (Luckhurst 79). For Luckhurst, trauma's 'shock impact … is anti-narrative, but it also generates the manic production of retrospective narratives that seek to explicate the trauma' (79). Indeed, this is one of the paradoxes of trauma: that it is so overwhelming and confusing as to pose major challenges to language and representation, and yet it is a widespread catalyst in the production of narrative. It seems that traumatic experiences can often compel survivors to compulsively repeat their experiences in creative forms. This is certainly the case with O'Brien, who has written on the Vietnam War repeatedly.

Trauma arguably manifests itself in similar ways in both the fictional and non-fictional work of authors who have suffered trauma—in ways that, this article suggests, might invalidate conventional ideas of genre. Authors such as Kathryn Harrison, Jeanette Winterson and Kurt Vonnegut have, like O'Brien, written repeatedly on traumatic events, exploring the same traumatic experiences again and again throughout individual autobiographically influenced creative texts. Harrison's memoir *The Kiss*, for example, pivots around the trauma of incest—autobiographical subject matter she also explores in her fiction. Similarly, Winterson's creative writing returns over and again to the trauma of her adoption, while Vonnegut's fiction is informed by his wartime experiences: the traumatic events of the Dresden bombings. Repetitions and gaps in the works of these creative writers resonate with the unconscious repetitions and absences in the testimonies of trauma survivors and their experiences of symptoms such as flashbacks, the re-experiencing of the traumatic past in the present, and patterns of dissociation.

The critic Amanda Wicks focuses on Vonnegut's traumatic wartime experiences, describing the repeated return to this material until the author seemingly finds his narrative or generic apotheosis within the science fiction genre. Wicks writes:

> Several of Vonnegut's earlier novels attest to his desire to write about Dresden; he integrates themes or subject matters of war, apocalypse, and biography into *Cat's Cradle* and *Mother Night* but still does not fully access the devastation of Dresden in the way he wanted. He finally lands upon the form and subject matter of science fiction, since the genre encompasses a wide range of topics that lie outside immediate human experience. (331)

According to trauma theory, such reworking of material constitutes evidence of a repetition compulsion characteristic of trauma. While the author attempts to access the past through a virtual space of fabulation—something that will be explored further in the following section—the traumatic events and experiences he wants to engage with prove elusive, and therefore repeat.

Harrison's published works include fiction and non-fiction in ways that seem to blend and blur generic classification, but which also repeat the same traumatic event. For example, the author's aforementioned memoir *The Kiss* was preceded by a novel titled *Thicker than Water*, in which a purportedly fictional young woman has an incestuous relationship with her biological father. This subject matter is explored again in *The Kiss*, in which arguably the most traumatic element of the author's experience is markedly absent from the narrative itself. Although Harrison depicts desire and moments of physical intimacy, she is elliptical in her descriptions of the consummation of her romantic relationship with her father, a fact she addresses explicitly in the memoir itself:

> In years to come, I won't be able to remember even one instance of our lying together. I'll have a composite, generic memory. I'll know that he was always on top and that I always

> lay still, as still as if I had, in truth, fallen from a great height. I'll remember such details as the colour of the carpet in a particular motel room, or the kind of tree outside the window. That he always wore his socks and that I wore whatever I could. I'll remember every tiny thing about him. I will be able to close my eyes and see the pattern of hair that grew on the backs of his hands, the mole on his cheek, the lines, each one of them, at the corners of his eyes. But I won't be able to remember what it felt like. No matter how hard I try, pushing myself to inhabit my past, I'll recoil from what will always seem impossible. (136–37)

While the book's sex scenes are not wholly omitted, as demonstrated in part by the passage above, they are evoked primarily as absent or fragmented memories and perceptions, enacting an obliqueness that Caruth describes as typical of the traumatic event and its aftermath—an absence that requires constant redressing.[1]

Further attesting to the ways in which repetitive compulsions are manifested in the creative practice of fiction writers are Winterson's novels. Winterson has authored numerous works of fiction that draw heavily—and often quite self-consciously—from her own lived experiences. Her first novel, *Oranges Are Not the Only Fruit*, was semi-autobiographical: supposedly a work of fiction, the book borrows extensively from Winterson's own childhood experiences in its depiction of a child who is adopted by evangelist parents. Later works include a number of magical realist texts, including *Sexing the Cherry*, a novel set in seventeenth-century London in which a gargantuan 'Dog Woman' adopts the baby she finds in the River Thames. Drawing on the 'Twelve Dancing Princesses' fairy-tale, the narrative frequently departs from historical realism and verisimilitude, and into fantasy and hyperbole.

Winterson writes in her more recent memoir, *Why Be Happy When You Can Be Normal*: 'When we write we offer the silence as much as the story. Words are the part of silence that can be spoken' (8). Emphasising the role of reading and writing in navigating and making sense or meaning out of difficult experiences—and channelling Adrienne Rich's poem 'Diving into the Wreck'—Winterson argues:

> I believe in fiction and the power of stories because that way we speak in tongues. We are not silenced. All of us, when in deep trauma, find we hesitate, we stammer; there are long pauses in our speech. The thing is stuck. We get our language back through the language of others. We can turn to the poem. We can open the book. Somebody has been there for us and deep-dived the words. (9)

In other words, Winterson's oeuvre enacts something typical of trauma: its compulsion to repeat. In an examination of trauma in Winterson's *The Stone Gods*, Onega writes that the protagonist 'Billie Crusoe's life story produces a strong effect of déjà vu, since it is the same deeply traumatic story of impossible love between an unwanted child and her red-haired teenage mother that Jeanette Winterson has been trying to tell all her life' (295). Even in an undecidedly science fiction narrative,[2] the author's real-life trauma remains repetitiously unresolved.

Similarly, while O'Brien's *Going After Cacciato* has been published as a work of fiction, as in Winterson's works, O'Brien's books suggestively function as fiction and autobiography, drawing from and performing these forms both simultaneously and in opposition to one another. *The Things They Carried*, for example, is prefaced with a blurb typical to many works of creative writing. It reads: 'This is a work of fiction. Except for a few details regarding the author's own life, all the incidents, names and characters are imaginary.' However, as Maria S. Bonn notes in an essay on the interplay of truth and imagination in O'Brien's representations of Vietnam, the character names and the details that follow, as well as the book's final notes and acknowledgements, complicate such a statement (12–14). The full names of purportedly fictional characters are listed in the book's dedication page, undermining the 'work of fiction' disclaimer.

In *The Things They Carried*, O'Brien self-consciously describes his obsessive return to the same scenes in ways that resonate with post-traumatic stress disorder:

> I'm forty-three years old, and a writer now, and the war has been over for a long while. Much of it is hard to remember. I sit at this typewriter and stare through my words and watch Kiowa sinking into the deep muck of a shit field, or Curt Lemon hanging in pieces from a tree, and as I write about these things, the remembering is turned into a kind of rehappening. Kiowa yells at me. Curt Lemon steps from the shade into bright sunlight, his face brown and shining, and then he soars into a tree. The bad stuff never stops happening: it lives on in its own dimension, replaying itself over and over. (32)

These virtual re-happenings are performed repeatedly throughout the text, even as O'Brien insists on an essential if impossible 'truth' in his fiction. Indeed, O'Brien frames the repetitiveness and unreliability of his text as evidence of its authenticity. In *The Things They Carried*, he claims: 'You can tell a true war story by the way it never seems to end' (73), and later: 'I want to tell you why story-truth is truer sometimes than happening truth' (179). Earlier in the book, the author explains:

> In any war story, but especially a true one, it's difficult to separate what happened from what seemed to happen. What seems to happen becomes its own happening and has to be told that way. … The pictures get jumbled; you tend to miss a lot. And then afterward, when you go to tell about it, there is always that surreal seemingness, which makes the story seem untrue, but which in fact represents the hard and exact truth as it *seemed*. (69–70)

Arva makes note of a compulsive and 'uncanny attention to detail' (271) in *Going After Cacciato*, a quality that also resonates with traumatic memory and its repetitive expression, and which is also similarly present throughout numerous books by O'Brien. *The Things They Carried*, for example, obsessively details burdens, both concrete and abstract, tangible and intangible, possible and impossible, intimate and vast in scale: 'They carried the sky. The whole atmosphere, they carried it, the humidity, the monsoons, the stink of fungus and decay, all of it, they carried gravity' (13). Pages later the refrain '[t]hey carried' is evoked again, as O'Brien writes: 'They carried all the emotional baggage of men who might die. Grief, terror, love, longing—these were intangibles, but the intangibles had their own mass and specific gravity, they had tangible weight' (17). These abstract burdens provide a contrast with the concrete and mundane items also listed, things 'largely determined by necessity': 'P-38 can openers, pocket knives, heat tabs, wristwatches, dog tags, mosquito repellent, chewing gum, candy, cigarettes, salt tablets, packets of Kool-Aid, lighters, matches, sewing kits, Military Payment Certificates, C rations, and two or three canteens of water' (3–4).

Sometimes, as in this excerpt and as noted by Arva, repetition works to obsessively recall and document detail. At other times, however, repetitions act as ambiguous icons of trauma, and the reader is positioned to experience their presence and placement in the text as akin to the traumatic symptoms experienced by the characters. That is, they function in a way that is unrelenting and without context, returning to haunt the narrative at unexpected times and places. In *Going After Cacciato*, for example, we note repeatedly that the character Billy Boy Watkins has died of fright. Announced at the beginning of the book's opening paragraph, this fact becomes something of an incantatory refrain, and its context and the scene to which it belongs chronologically is revealed gradually.

As suggested by the ways in which the repetitious and unresolvable nature of the trauma often segues with the fantastical, a number of trauma narratives use magical realist strategies to depict trauma, representing the magical alongside the historically real. Indeed, magical realism is a literary mode that reveals some of the key elements of trauma narratives, including repetition, along with disruptions to linear modes of storytelling, dissociation and the potent presence of the unspoken, which arrives and persists by way of ghosts and other fabulated forms. Testimony and magical realism can thus arguably be seen to converge when it comes to representing trauma, testing the limits of these genres.

Trauma and fantasy

Given trauma's challenges to the possibility of realist representation, a number of critics have noted the ways in which genres that embrace the fantastic read as especially attuned to the difficulties of conveying the incomprehensible. Francisco Collado-Rodríguez, for example, describes the gothic genre's congruence with trauma narratives, noting that 'trauma narratives center on the representation of the incomprehensible, the irrational, and the forgotten or never memorized traumatizing event that blocks the victim's recovery', all of which, he contends, 'also occurs in the case of gothic literature' (624). Vonnegut's iconic text provides an interesting example. As noted, the novel mimics traumatic repetitions, and, in Luckhurst's words, 'explodes the possibility of living in sequential time … [as] a post-traumatic survivor' (Luckhurst 161). Attending to the fantastic elements and science fiction conventions in Vonnegut's novel, Wicks contends that '[s]cience fiction offers new ground for trauma fiction' (338). She argues:

> Where more traditional literary genres may limit the ability to narrate and transmit the experience of trauma, science fiction moves closer to understanding that which resists comprehension. Steeped in cognitive estrangement, which alternates between the familiar and the unfamiliar, the genre produces unique insights into the acts of both experiencing and narrating trauma. (338)

Arva puts forward a similar argument about magical realism, proposing a unique connection between representations of trauma and extremity, and the stylistic devices of magical realism. The magical realist image, Arva argues, is 'capable of bringing the pain and horror home into the readers' affective world; while it might not need to explain the unspeakable … it can certainly make it felt and re-experienced in a vicarious way' (9). In this way, Arva suggests, magical realism's departure from a tangible and recognisable reality into the fantastical or heavily metaphorical simulates the traumatic experience as it was felt at the time —or as it is remembered and grasped by survivors later, through memories which are frequently fragmentary and marked by absence, intensity or distortion. Arva refers to the 'intrinsically uncanny reality of traumatic events—histories that were and were not at the same time' (6).

Arva thus conceives of magical realism as self-conscious in its refusal to separate the realistic and the fantastic—a position that resonates with existing scholarship. However, whereas critics such as Christopher Warnes, Maggie Ann Bowers and Wendy B. Faris argue that magical realism challenges the realism and authority of patriarchal or colonial discourses, Arva focuses on the ways in which magical realist texts challenge realism's ability to represent trauma. 'By virtue of its subversive character,' Arva argues, 'magical realism foregrounds, somewhat paradoxically, the falsehood of its fantastic imagery exactly in order to expose the falsehood—and traumatic absence—of the reality it has proposed to represent' (62). Here Arva suggests that by acknowledging the unknowability of violent and traumatic historical events, and simultaneously undermining the possibility of authoritative representation, magical realism paradoxically provides an accurate or faithful mode of expression, in which the reader is positioned to acknowledge the comprehensible and incomprehensible all at once, recognising that which is absent along with that which is (shockingly) present.

Notably employing the language of metaphor to illustrate his contention, Arva argues that '[o]ne must understand the magical realist universe not as a flight from reality but as a flight simulator, an artificial world within the real world, meant to prepare us for a better grasp of it' (108–9). By disrupting narrative patterns, blending the palpably real with unreal or fantastic events, and affecting a sense of disassociation or defamiliarisation, magical realist narratives position the reader to experience events of extremity as they are perceived and remembered by survivors. To force an incoherent experience and its subsequently fragmented memory into a coherent realistic narrative would be, by contrast, to construct something false and contrived—something counterfeit to the original experience. What Arva contends is that magical realism offers a more realistic mode of representing that which cannot be represented.

In fact, Arva goes as far as to argue that '[m]agical realism writes the silence that trauma keeps referring to, and converts it into history' (23). For Arva, magical realism allows for this listening and seeing via affect: what he describes as a '*felt* reality', something paradoxically 'unreal but true' (112). That is, magical realism highlights the experience's factitiousness or lack even as it engages the reader with the potency of the representation.

In discussing the unreal qualities of Winterson's magical realist fiction, Onega similarly suggests that 'the value of experimental fiction would lie not so much in its capacity to represent trauma or its aporias, but in its capacity to shock the reader out of habituation and numbing and into affective participation and sensorial understanding of trauma' (269). While Onega refers to 'experimental fiction' as opposed to magical realism, and to the postmodern novel *The Stone Gods* rather than the more conventionally magical realist *Sexing the Cherry*, she nonetheless indicates a line of interest similar to Arva's notions of simulation and affect. For Arva, what magical realism specifically offers is pure or true 'sensation'. Through the author's use of language and the imaginative experience offered by the text, the reader is positioned to see and feel, without necessarily understanding or being able to contextualise, the traumatic event and its impact.

While this approach might provoke ethical concerns about the possibility for historical denial, Arva insists that 'the relationship between the real and the imaginary is not the same as the one between truth and untruth' (75). He suggests that when an experience is beyond typical or traditional means of realistic representation, authors are required to be exceptionally creative in order to give expression to such events—and to access and provide a version of reality that *feels* true.

The pattern of 'felt' trauma described by Arva is one present in O'Brien's *Going After Cacciato*—indeed, Arva dedicates a chapter of his study to O'Brien's book, pairing it with Günter Grass's *The Tin Drum* in an examination of what he calls 'shock chronotopes', borrowing from Mikhail Bakhtin's definition of the 'chronotope' to articulate a 'traumatic time-space' (Arva 5). For Arva:

> Imagination, and especially the traumatic imagination, is an activity by which the human consciousness translates an unspeakable state ... into a readable image; it is the process by which shock chronotopes become artistic chronotopes. The traumatic imagination uses the sublimative power of language in order to turn that which resists representation into a new and tangible reality. (84)

Jenny Edkins also evokes what she calls 'trauma time', contrasting such disruptive chronotopes with linear versions of history and progress and the ways in which they serve hegemony. She writes:

> In the linear time of the standard political processes, which is the time associated with the continuance of the nation-state, events that happen are part of a well-known and widely accepted story. What happens fits into a pattern. ... In trauma time, in contrast, we have a disruption of this linearity. Something happens that doesn't fit, that is unexpected – or that happens in an unexpected way. It doesn't fit the story we already have, but demands that we invent a new account, one that will produce a place for what has happened and make it meaningful. Until this new story is produced we quite literally do not know what has happened: we cannot say what it was, it doesn't fit the script – we only know that 'something happened'. (xiv)

For Arva, magical realism allows for the silent 'something happened' of trauma to be made present and meaningful. Such an envisioning of magical realism will be further explored in the section to follow, via a close reading of O'Brien's work and the shock chronotope it mobilises: the author's traumatic experiences during the Vietnam War.

As Stefania Ciocia writes, 'O'Brien legitimizes and practises a certain narrative embellishment of factual reality' and, in so doing, can 'salvage, and then communicate, the exact intensity of the original impact of the narrated events on those who experienced them, either first-hand, or even only as powerful stories' (5). Tobey C. Herzog, emphasising the authenticity of the unreal in O'Brien's depictions of his traumatic experiences of the Vietnam War, writes that to omit the presence of 'memories, fantasies, and dreams ... would be to impose an artificiality on his works that would destroy their realism' (26). O'Brien himself suggests, both in metafictional creative passages and in the claims he makes during interviews, that it is the impact of trauma that destroys narrative realism —and, in line with Caruth's position, that this needs to be acknowledged in trauma narratives. At the same time, however, the author is quick to caution the reader against accepting the veracity of his narratives. In *The Things They Carried*, he writes: 'In many cases a true war story cannot be believed. If you believe it, be skeptical' (70). Such a statement is at odds with earlier narrative claims, including lines such as: 'This is true' and 'It's all exactly true' (67 & 69). This returns us to the idea noted earlier, which is that trauma narratives typically merge and even refuse to reconcile elements of autobiography and fiction, thus claims pertaining to authenticity are complicated by artfulness.

'A true war story that never happened': Tim O'Brien's magical realist life writing

At every turn, *Going After Cacciato* emphasises the imbrications of fantasy in the experience of trauma. The initial departure from the 'reality' of war is spurred by Private Cacciato, who deserts the war with the intention of walking to Paris—an ironic destination, given Vietnam's colonial past, and perhaps moreover the fact that the Vietnam War is itself an imperialist war. The remaining members of his squad, including O'Brien's protagonist, Paul Berlin, are sent after him, and it is during the pursuit that the destination of Paris becomes Berlin's dissociative fantasy. In actuality, the squad turns back after Cacciato makes it across the border into Laos, and throughout the novel the reader is repeatedly teased by reminders that the ensuing journey to Paris is a product of Berlin's imagination.

While distinctions between the real and imagined are often unclear, Berlin's propensity for dissociation is made apparent early in the narrative. Shortly after the characters set out after Cacciato, we see him absorbed in an undertaking of imagination and fabulation. O'Brien writes:

> Paul Berlin sat alone, playing solitaire in the style of Las Vegas. Pretending ways to spend his earnings. Travel, expensive hotels, tips for everyone. Wine and song on white terraces, fountains blowing colored water. Pretending was his best trick to forget the war. (10)

Although the lieutenant interrupts the fantasy in the following line, signalling the end of Berlin's daydreams, the protagonist continues to escape via his imagination. While Berlin realises that the squad can only follow Cacciato so far, he would like to keep up the pursuit. As the group walk through the high country, we read: 'He liked the silence. He liked the feel of motion, one leg and then the next. No fears of ambush, no tapping sounds in the brush. The sky was empty. He liked this' (16). Thus the voyage to Paris that ensues is flagged as a fantasy of wish-fulfilment: 'Walking away, it was something fine to think about. Even if it had to end, there was still the pleasure of pretending it might go on forever: step by step, a mile, ten miles, two hundred, eight thousand.'

The unreal road to Paris that ensues is littered with Cacciato's 'empty ration cans, bits of bread, a belt of gold-cased ammo dangling from a shrub, a leaking canteen, candy wrappers, worn rope' (18). There are also more comical and even more improbable clues, such as the map that Cacciato has tacked to a log on the road, which the characters find marked with a caution inside 'a precisely drawn circle'. Berlin sees that: 'Within the circle, in red, were two smaller circles, between them an even smaller circle, and beneath them a big banana smile. A round happy face. Underneath it, in printed block letters, was a warning: LOOK OUT, THERE'S A HOLE IN THE ROAD' (73). Following this, Berlin's fantastical escape road to Paris becomes increasingly unstable, collapsing into tunnels and holes, and generating a state of hallucinatory confusion resonant of Lewis Carroll's *Alice's Adventures in Wonderland*. At one point Berlin finds himself falling through the earth alongside members of his squad and the trio of refugees—including their buffalo and cart—with whom they have found themselves travelling:

> So down and down, pinwheeling freestyle through the dark … Far below he could make out the dim tumbling outline of the buffalo and slat-cart, the two old aunties still perched backward at the rear. He heard them howling. Then they were gone. His lungs ached. The blood

> stopped in his veins, his eyes burned, his brain plunged faster than his stomach. The hole kept opening. Deep and narrow, lit by torches that sped past like shooting stars, red eyes twinkling along sheer rockface, down and down. (82)

The falling scene continues: 'So down and down, pinwheeling freestyle through the dark' (82); 'Down and down. Or up and up, it was impossible to know' (99).

Indeed, impossibility and its double, possibility, characterise the novel. As noted, Berlin, from the outset, chooses fiction over reality, and his fantasies seem to constitute the plot's most significant events—its climaxes and conflicts. Just as Berlin's propensity for dissociative fantasy is made apparent early on in the narrative, so too is the news of Cacciato's desertion a signal of the kind of imaginative dissociation that will come to characterise Berlin's experience of things:

> He imagined it. He imagined the many dangers of the march: treachery and deceit at every turn, disease, thirst, jungle beasts crouching in ambush; but, yes, he also imagined the good times ahead, the sting of aloneness, the great new quiet, new leanness and knowledge and wisdom. The rains would end. The trails would go dry, the sun would show, and, yes, there would be a changing foliage and seasons and great expanses of silence, and songs, and pretty girls sleeping in straw huts, and, where the road ended, Paris. (23)

This phantasmic synopsis is fleshed out in the rest of what follows. Indeed, the passage quite accurately indicates the imagined action to come. Towards the end of the novel, the 'songs' occur with the arrival of Christmas and a celebratory mood: 'They spent the night around the tree. Drinking, making awkward talk, they trimmed the spruce with medals and strings and grenades and candles. Later they tried a few carols, then smoked the last of Oscar's precious dope' (182). The 'pretty girls' take shape in Berlin's love interest, Sarkin Aung Wan, the youngest woman of the refugee trio who is, in fact, described as 'just a creature of his own making—blink and she was gone' (202). A figment of Berlin's chimerical reality, Sarkin Aung Wan nonetheless has a physical, sensory form. She has sparkling gold hoop earrings, and smells of 'soap and joss sticks' (53). Her damp hair is described as 'like seaweed on his leg'; she is '[e]verything so soft' (114).

Paris, when Berlin arrives there, is similarly ambiguous; it 'comes like a ghost', but is also rendered with a visual intensity:

> A swirling skyline. Jagged Gothic towers touching the clouds. Bridges and billboards and a swirl of concrete and bricked things appearing and snapping away like magic - a house with painted shutters, a bakery, a man walking his dog, warehouses, gleaming puddles, streets and parks and umbrellas. (290)

When Cacciato had first deserted his post the lieutenant had joked: 'gay Paree—bare ass and Frogs everywhere, the Follies Brassiere' (5). However, when Berlin reaches his imagined and idealised destination, it is no longer a mere caricature; instead the sense of place is impossibly, self-consciously real: 'He feels the wind—it's real. He licks rain from his lips. Real rain—wet and real. If you can imagine it, he tells himself, it's always real. Even peace, even Paris—sure, it's real' (291).

However, Berlin's imagination cannot successfully transport him somewhere completely removed from the conflict of his material circumstances. Even his fantasies are haunted by violence and extremity. In Tehran, which is among the cities the squad pass through on their journey, the characters are captured and sentenced to death for their alleged desertion—and in truth, the protagonist has deserted the war by way of his imagination. O'Brien writes: 'Things were out of control. Gone haywire. You could run but you couldn't outrun the consequences of running. Not even in imagination' (226). Ludicrously, the group escapes with the help of Cacciato. Their getaway is outrageous and factitious: lit by 'garish colours and searchlights swaying through the night and a city full of sirens—just a cartoon' (244). However, at the fantastic climax, in which the squad finally closes in on Cacciato in Paris, Berlin is dropped back into reality. He returns to Vietnam, where his squad is truly close to apprehending Cacciato at the border, and finds himself overwhelmed by past traumas. Berlin shoots his weapon too soon, and literally shits himself: 'a wet releasing feeling' (331). Cacciato gets away, and the characters are, presumably, returned to the traumatic realities of the war. In this way, the text highlights the awful inescapability of the war, which has no glamour. The exciting allure of the novel's 'thriller' conventions is subverted by the actual conclusion: the departure into fantasy is followed by a brutal return to reality.

Although *Going After Cacciato* might be more explicit in its departures into fantasy, the short-story collection *The Things They Carried* also comprises elements of fantasy, and the spaces of memory and imagination provide frequent departures from, or even overtake, the characters' immediate narrative settings. The collection also, and even more patently, blurs the boundaries of life writing and magical realism when it comes to representations of trauma. As in *Going After Cacciato*, the stories' spaces are often ghostlike, but also intensely rendered realities that work to signal the characters' proclivity for dissociation. The stories also destabilise one another, as we see in the following passage, which recalls key details from earlier scenes in the book:

> No trail junction. No baby buffalo. No vines or moss or white blossoms. Beginning to end ... it's all made up. Every goddamn detail – the mountains and the river and especially that poor dumb baby buffalo. None of it happened. *None* of it. And even if it did happen, it didn't happen in the mountains, it happened in this little village on the Batangan Peninsula. (80)

Indeed, the stories often provide metafictional commentary on their counterparts. We see this, for instance, through Jimmy Cross, the main focalised character in the first and titular story, who tells the narrator character of the following story, a writer named Tim O'Brien: 'Make me out to be a good guy, okay? Brave and handsome and all that stuff. Best platoon leader ever' (27).

Repetition also characterises the stories at the level of language itself. There is the repetitive description of the man O'Brien's eponymously named character killed during the war: the image of a 'star-shaped hole' indicative of this man's facial wound spans numerous pages (121, 122, 125 & 130). The scene is otherwise almost idyllic and thus incongruent with the violent imagery. There are 'small white flowers shaped like bells' (123) and sunlight that 'sparkle[s]' (124). There is 'a butterfly on [the man's] chin', 'making its way along [his] forehead' (121 & 123).

The stories also draw the reader's attention to the traumatically occluded: experiences which might be marginalised by trauma's impact, or by the social invalidation of certain traumatic events. In 'Speaking of Courage', the story's narrator conveys 'the terrible killing power of [a] shit field' (158), and recalls how, upon the soldiers' return home after the war, 'nobody in town wanted to know about the terrible stink' (148). Indeed, this is what makes the text's uncertain elements politically charged, and therefore in need of attending to as more than the traumatic unknown that ought to be preserved as such.

Conclusion

Magical realist literature frequently calls attention to limits: it tests the bounds of plausibility and veracity, language and representation, and confronts the limits of, and absences in, hegemonic historical discourse. In doing so, the mode frequently enacts Linda Hutcheon's theory of historiographic metafiction, a 'seriously ironic parody' comprised of 'the parodic reworking of the textual past of both the "world" and literature' (4). Many of the texts that constitute the genre of life writing also share this impulse or methodology. As Clara Juncker contends of O'Brien's work, the author 'focuses on the process of writing Vietnam as much as on experiencing it' (111). In this respect, magical realism might pose a provocative challenge to or reconceptualization of the category of life writing, in which lived experience, traumascapes, and the imagined intersect within powerful and dynamic textual spaces.

Notes

1. Of note is that, as Luckhurst points out, '[k]ey scenes [from *Thicker Than Water*] reappeared word for word [in *The Kiss*], but the novel also contained sexual details that had been occluded in the memoir' (134–44). These kinds of repetitions and discrepancies might have provided fuel for Harrison's detractors: Luckhurst describes how '*The Kiss* ... provoked a campaign of press vilification and prompted several commentators to observe that the 1990s memoir boom had now found its shipwreck' (143). Notably, such attacks are typically gendered, affirming that notions of truth and reality—and of violence and trauma—are dominated by patriarchy. As Gilmore points out, 'first-person accounts of trauma by women ... are likely to be doubted, not only when they bring forward accounts of sexual trauma but also because their self-representation already is at odds with the account the representative man would produce' (23). Of interest to this article, however, are the ways in which the unresolved trauma repeats in varying oblique and explicit ways.
2. While critics and reviewers have struggled to place *The Stone Gods* as science fiction or otherwise, Winterson has argued that such labels are meaningless. On her website, the author writes: 'I can't see the point of labelling a book like a pre-packed supermarket meal. There are books worth reading and books not worth reading. That's all.'

Disclosure statement

No potential conflict of interest was reported by the author.

References

Arva, Eugene. *The Traumatic Imagination: Histories of Violence in Magical Realist Fiction*. Amherst, NY: Cambria Press, 2011.
Bonn, Maria S. "Can Stories Save Us? Tim O'Brien and the Efficacy of the Text." *Critique: Studies in Contemporary Fiction* 36.1 (1994): 2–15.
Bowers, Maggie Ann. *Magic(al) Realism*. London: Routledge, 2004.
Carroll, Lewis. *Alice's Adventures in Wonderland*. London: Penguin Books, 2006 (1865).
Caruth, Cathy, ed. *Trauma: Explorations in Memory*. Baltimore: Johns Hopkins University Press, 1995.
Caruth, Cathy. *Unclaimed Experience: Trauma, Narrative and History*. Baltimore: Johns Hopkins University Press, 1996.
Ciocia, Stefania. *Vietnam and Beyond: Tim O'Brien and the Power of Storytelling*. Liverpool: Liverpool University Press, 2012.
Collado-Rodríguez, Francisco. "Textual Unreliability, Trauma, and the Fantastic in Chuck Palahniuk's *Lullaby*." *Studies in the Novel* 45.4 (2014): 620–637.
Edkins, Jenny. *Trauma and the Memory of Politics*. Cambridge: Cambridge University Press, 2003.
Faris, Wendy B. *Ordinary Enchantments: Magical Realism and the Remystification of Narrative*. Nashville: Vanderbilt University Press, 2004.
Gilmore, Leigh. *The Limits of Autobiography: Trauma and Testimony*. Ithaca: Cornell University Press, 2001.
Harrison, Kathryn. *Thicker Than Water*. New York: Random House, 1992.
Harrison, Kathryn. *The Kiss*. London: Fourth Estate, 1997.
Herman, Judith Lewis. *Trauma and Recovery: from Domestic Abuse to Political Terror*. London: Pandora, 2001 (1992).
Herzog, Tobey C. *Tim O'Brien*. New York: Twayne Publishers, 1997.
Hutcheon, Linda. "Historiographic Metafiction: Parody and the Intertextuality of History." *Intertextuality and Contemporary American Fiction*. Ed. Patrick O'Donnell and Robert Con Davis. Baltimore: The Johns Hopkins UP, 1989. 3–32.
Juncker, Clara. "Not a Story to Pass On?: Tim O'Brien's Vietnam." *Transnational America: Contours of Modern U.S. Culture*. Ed. Russell Duncan and Clara Juncker. Copenhagen, Denmark: Museum Tusculanum, 2004. 111–124.
Luckhurst, Roger. *The Trauma Question*. New York: Routledge, 2008.
Luckhurst, Roger. "Future Shock: Science Fiction and the Trauma Paradigm." *The Future of Trauma Theory: Contemporary Literary and Cultural Criticism*. Ed. Gert Buelens, Sam Durrant, and Robert Eaglestone. Hoboken, NJ: Taylor and Francis, 2014. 157–167.
O'Brien, Tim. *The Things They Carried*. London: Flamingo (Harper Collins), 1991 (1990).
O'Brien, Tim. *Going After Cacciato*. New York: Broadway Books, 1999 (1978).
Onega, Susana. "The Trauma Paradigm and the Ethics of Affect in Jeanette Winterson's *The Stone Gods*." *Ethics and Trauma in Contemporary British Fiction*. Ed. Susana Onega and Jean-Michel Ganteau. New York: Rodopi, 2011. 265–298.
Pederson, Joshua. "Speak, Trauma: Toward a Revised Understanding of Literary Trauma Theory." *Narrative* 22.3 (2014): 333–353.
Rich, Adrienne. *Diving into the Wreck: Poems 1971–1972*. New York: Norton, 1973.
Takolander, Maria, and Jo Langdon. "Shifting the 'Vantage Point' to Women: Reconceptualizing Magical Realism and Trauma." *Critique: Studies in Contemporary Fiction* 58.1 (2017): 41–52.
Vonnegut, Kurt. *Slaughterhouse-five*. New York: Dell Publishing (Random House), 1991 (1969).
Warnes, Christopher. *Magical Realism and the Postcolonial Novel: Between Faith and Irreverence*. Basingstoke: Palgrave Macmillan, 2009.
Whitehead, Anne. *Trauma Fiction*. Edinburgh: Edinburgh University Press, 2004.
Whitlock, Gillian, and Kate Douglas. *Trauma Texts*. London: Routledge, 2009.
Wicks, Amanda. "'All This Happened, More or Less': The Science Fiction of Trauma in *Slaughterhouse-five*." *Critique: Studies in Contemporary Fiction* 55.3 (2014): 329–340.
Winterson, Jeanette. *Oranges Are Not the Only Fruit*. London: Bloomsbury Publishing, 1991 (1985).

Winterson, Jeanette. *The Stone Gods*. London: Hamish Hamilton, 2007.
Winterson, Jeanette. *Sexing the Cherry*. London: Vintage, 2009 (1989).
Winterson, Jeanette. *Why Be Happy When You Can Be Normal?* London: Jonathan Cape, 2011.
Winterson, Jeanette. "The Stone Gods". *Jeanette Winterson*. 2013, http://www.jeanettewinterson.com/book/the-stone-gods/.

Forms of Resistance: Uses of Memoir, Theory, and Fiction in Trans Life Writing

Juliet Jacques

ABSTRACT

This article examines the forms of writing that transgender (hereafter 'trans') people in North America, the United Kingdom and Australia have used to convey their experiences to a wider public, since the first sex reassignment surgeries were performed during the inter-war period. I will discuss how transsexual people first used memoir to counter sensationalistic mass media coverage, and then how feminist critiques of their works led to 'post-transsexual' theory, which deconstructed the conventions and clichés that the transition memoir genre had developed. These theorists suggested new ways to document gender-variant lives beyond a generalised desire for public acceptance and in response to social concerns: the policies of the gender identity clinics, who decided who could access medical services, and on what terms, especially the demand that patients 'pass' in their acquired genders and hide their histories; and the exclusion of trans people from feminist spaces, and gay/lesbian politics. The article concludes with a consideration of my own autobiographical writing, in a newspaper blog and a memoir, and how my practice responded to and attempted to change how trans people were expected to write about their lives.

This article examines the forms of writing that transgender (hereafter 'trans') people in North America, the United Kingdom and Australia have used to convey their experiences to a wider public, since the first sex reassignment surgeries were performed during the inter-war period. I will discuss how transsexual people first used memoir to counter sensationalistic mass media coverage, and then how feminist critiques of their works led to 'post-transsexual' theory, which deconstructed the conventions and clichés that the transition memoir *genre* had developed. These theorists suggested new ways to document gender-variant lives beyond a generalised desire for public acceptance and in response to social concerns: the policies of the gender identity clinics, who decided who could access medical services, and on what terms, especially the demand that patients 'pass' in their acquired genders and hide their histories; and the exclusion of trans people from feminist spaces, and gay/lesbian politics. Post-transsexual theorists' texts often drew heavily on their own experiences: thus, they ensured that directly autobiographical writing would remain the dominant mode of trans discourse, even as they rejected the established structures of memoir. Indeed, the most interesting, formally inventive theory included plenty of autobiographical material, looking at how transphobia in places from the medical establishment to mass media impacted on our actual lives – a line that could be traced from Bornstein via Julia Serano's influential *Whipping Girl: A Transsexual Woman on Sexism and the Scapegoating of Femininity* (2007) through to *Testo Junkie: Sex, Drugs and Biopolitics in the Pharmacopornographic Era* (2013) by Beatriz (later Paul B.) Preciado. Here, I will explore how trans-identified authors have used fiction, or a blurring of boundaries between autobiography and fiction, to resist some of the structural and social limits of trans life writing, and suggest ways in which this formal exploration may be more rewarding, both aesthetically and politically. I will then discuss representations of trans people in mainstream media in the United Kingdom, ways in which I used life writing to challenge the terms of that representation, and how authors might use a greater variety of fictional (in addition to or instead of autobiographical) techniques to explore a wider range of gender positions.

The first sex reassignment surgeries grew out of late nineteenth and early twentieth century sexology – and specifically its use of gender-variant life stories. Berlin-based sexologist Magnus Hirschfeld thought that rather than being inherently linked to homosexuality (as his London counterpart Havelock Ellis suggested, with his theory of 'sexual inversion'), cross-dressing represented an 'independent complex' that could not 'be ordered according to recognised models' (28–29). To test this, Hirschfeld interviewed numerous people who wanted to dress as – or *be* – the opposite sex. He based his conclusion that gender identity was a separate issue to sexual orientation on their testimony, which formed a crucial aspect of his influential survey, *The Transvestites: The Erotic Drive to Cross-Dress* (1910).

After the First World War, Hirschfeld continued his research, which led to the separation of 'transvestite' and 'transsexual' into distinct categories. He supervised some of the surgeries performed on Danish artist Lili Elbe, who died in 1931 after a failed attempt to transplant a uterus into her body. Her life story, *Man into Woman,* was published posthumously in 1933 – the first autobiographical account of a transsexual life. Intriguingly, this text played with the conventions of the transition memoir even before they were set: this was not a linear story, nor was Elbe its sole author. Although the book incorporated Elbe's diaries and letters, and conversations with the 'editor' Niels Hoyer, who ordered her fragments, its protagonist used the name 'Andreas Sparre' (rather than Elbe, or Einar Wegener, as she was previously known), while Hoyer was a pseudonym for the German writer Ernst Ludwig Jacobson (Harrod, 'The tragic true story'). Later, I will discuss Philippe Lejeune's 'pact', which said that 'identity between author's and narrator-protagonist's name is the primary requisite of autobiography' (Marcus 193), but these pseudonyms were less likely an attempt to play with the (unspoken) laws of life writing, and more probably (in Hoyer's case) a self-protective measure, used in the awareness that the one of first organisations the Nazis had attacked on taking power in January 1933 was Hirschfeld's Institute for Sexual Science.

After the Second World War, research into gender variance began again – in the United States, where displaced German sexologists such as Harry Benjamin had settled. In December 1952, transsexual people became global news, when the *New York Daily Times* featured Christine Jorgensen under the headline 'Ex-GI Becomes Blonde Beauty'. Jorgensen became the subject of frenzied media coverage, as did models and actors such as April Ashley or Caroline Cossey, who were outed by the British tabloids. Despite (or maybe *because*) of this, many more people came out over the following decades, as transsexual people went from being isolated examples of a scientific possibility to a group that threatened established categories of male and female – positing a threat both to conservative gender roles and to a burgeoning second-wave feminist movement whose tactic of creating 'women-only' spaces relied upon strict gender demarcation. During this time, with growing public interest in the subject, numerous transition memoirs were published, mostly by transsexual women (including Jorgensen and Ashley). Consequently, the transition memoir became a genre, with recognisable clichés and conventions – not all of which sat comfortably with feminist or gender-variant readers.

Many of these conventions were codified in its most famous exponent, *Conundrum* (1974) by Jan Morris, who differed from most authors of similar books in that, being an acclaimed travel writer, she was known for something besides transition (and so could avoid being typecast as a 'transsexual author'). In 1972, she came out as transsexual; knowing that this would be of interest to the British press, she took control of her story. The resultant book was set in a heterosexual and cisgender context, positioning her transsexual impulse as a problem to be weighed against the possibility of losing her family, social circles or career, opening with the realisation that 'I [*was*] born into the wrong body, and should really be a girl' (Morris 79) and climaxing with the surgery, undertaken with Dr Georges Burou in Casablanca.

In 1987, the US artist, activist and academic Sandy Stone wrote 'The *Empire* Strikes Back: A Posttransexual Manifesto', which became influential after being distributed on early online networks. This was a reply to the academic/activist Janice Raymond's notorious *The Transsexual Empire: The Making of the She-Male* (1979), which cast male-to-female transsexuality as a plot to infiltrate the feminist movement's spaces, prompted in part by her resentment of Stone's employment at the women-only collective Olivia Records. Addressing Stone, Raymond wrote that 'transsexually constructed lesbian-feminist[s]' were able to '[insert] themselves into the positions of importance and/or performance in the feminist community' (133) by duping people into accepting them as women, blaming Stone for creative 'divisiveness' in that community; this, implied Raymond, was an ever greater crime than that committed by most transsexual women, who 'attempt to possess women in a bodily sense while acting out the images into which men have molded women' (that is, conforming to patriarchal stereotypes of femininity) (132).

Stone's manifesto did not open with Raymond, however, but with critiques of Morris and Elbe – this implicitly set up a dichotomy between memoir as something written for 'outsiders' while theory was *for* the trans community. Here, Stone examined how they and other memoirists provided little sense of continuity between male and female, casting their surgery as the moment when they became women. 'No wonder [*that*] feminist theories have been suspicious', wrote Stone, highlighting conflations of physical sex with learned gender roles, such as the point in *Man into Woman* that implausibly described Elbe's handwriting becoming 'a woman's script' after surgery (225). Stone asked the vital question of who these texts were *for*, as the gender identity clinics that handled patients did not consider them as reliable insights into the transsexual condition, and continued to tell transsexual people to 'erase [*themselves*], to fade into the "normal" population as soon as possible' (230).

Stone identified the imperative to 'pass' as the main barrier to honest transsexual life writing – in particular, 'passing' made it impossible to counter the feminist argument that transsexual people conformed to 'stereotypical behaviours' that, as sociologist Carol Riddell had argued in *Divided Sisterhood* (a response to Raymond published in 1980), were 'prescribed by patriarchy for either sex' (146). Stone took up this point, arguing that 'Transsexuals who pass [*in their acquired gender*] ignore the fact that by creating totalized, monistic identities … they have foreclosed the possibility of authentic relationships' (132) (an activity 'familiar to the person of color whose skin is light enough to pass as white, or to the closet gay or lesbian'.) Instead, argued Stone, such people enter the discourse around transsexuality, conceiving themselves 'not as a class or problematic "third gender"' but 'as a *genre* – a set of embodied texts whose potential for *productive* disruption of structured sexualities and spectra of desire has yet to be explored' (231). This inspired a generation of North American activist-theorists, who focused more on the social challenges of gender-variant life; those who did not abandon the conventional memoir format certainly adapted it freely to suit their political purposes.

One problem with transition memoirs was that they created an impression of people being focused more on themselves than any wider community. This was a structural issue in a genre that necessarily concentrated on individual experience, exacerbated by its long-standing role in countering lurid media coverage of transsexual lives by placing them within 'respectable' heterosexual and gender-normative contexts. Edited volumes of life stories, such as *Trans-X-U-All* (1997) and *Finding the Real Me* (2003), both collated by Tracie O'Keefe and Katrina Fox, aimed to create a sense of diversity, including people of colour, female-to-male, non-binary or 'genderqueer', and/or working-class people who diverged from the familiar story – that of the transsexual woman making a clear move from 'male' to 'female' via hormones and surgery (which, in the US, were expensive to access). Such collections could also circumvent the established narrative by focusing on a specific issue, such as the relationship between gender and sexuality, explored by authors such as Greta Christina, David Harrison, and D. Travers in *PoMoSexuals: Challenging Assumptions about Gender and Sexuality* (Queen and Schimel).

The playwright and performance artist Kate Bornstein contributed to *PoMoSexuals*, having made her name with *My Gender Workbook* and *Gender Outlaw* (1994), a hybrid text that presented her performance scripts and plays alongside chapters that looked at how the media and wider prejudice had shaped her life and work. She lamented a limit placed upon trans life writing by audiences, or by editors' and publishers' *expectations* of their audiences: that 'up until the last few years, all we'd be able to write *and get published* were our autobiographies … the romantic stuff which set in stone our image as long-suffering, not the challenging stuff.' (Bornstein 12–13) In response, Bornstein suggested that authors create new metaphors that would generate a 'transgendered writing style' (Bornstein 4), and change the terms of discussion. Like Sandy Stone, Bornstein emphasised the need for open dialogue *led* by trans people, rather than 'passing'. 'Silence does equal death,' Bornstein wrote, referencing Act Up's slogan about the HIV/AIDS crisis. 'That principle applies to any culturally-mandated silence', which Bornstein described as 'the therapeutic lie' (recommended by the clinics, whose purpose was apparently to look after the mental wellbeing of patients) 'that eventually causes us to go mad'(-Bornstein 94).

Gender Outlaw followed Stone's suggestion that 'posttranssexual' people mix genres, both in the author's refusal to identify as either male or female and in its structure. Alternate chapters used autobiographical material to make theoretical points, opening with one about how hard it had been for Bornstein to realise and express a 'non-traditional gender identity' in a society where anyone who did was ridiculed and monstered in the media (Bornstein 8). From there, she experimented with numerous methods of representation: transcripts of genuine interviews and invented ones with interrogators who had impersonal names such as 'Issues'; quiz questions for readers; fragments of poetry; anecdotes about how Bornstein's gender presentation was received, from school camping trips to TV chat shows; film criticism; reflections on texts such as Suzanne Kessler and Wendy McKenna's *Gender: An Ethnomethodological Approach* (1978); and scripts in which Bornstein's ideas about gender identity and society were used in semi-fictional or performative contexts. (That said, it was always clear which sections were theoretical, which were autobiographical and which used some creative license or artifice.)

Eventually, Bornstein did write a more conventional memoir – *A Queer and Pleasant Danger* (Bornstein), billed on its cover as 'The true story of a nice Jewish boy who joins the Church of Scientology and leaves twelve years later to become the lovely lady she is today', and notable for its extensive description of Bornstein's sexual experiences, before and after transition, rarely found in preceding memoirs. Perhaps, after two decades in which she and others had made memoir less dominant over trans writing, Bornstein felt able to revisit the form with some of the recent critiques of it in mind, as well as using it to cover other aspects of her life.

*

Stone Butch Blues: A Novel, by Bornstein's contemporary Leslie Feinberg (1949–2014), also expressed a position between male and female, but pushed even harder against the limits of life writing, blurring the lines between 'fiction' and 'memoir' in ways that even *Gender Outlaw* did not. Published by Firebrand in 1993, it won a Lambda Literary Award (Small Press book) and the American Library Association's Lesbian/Gay Book Award (Fiction). Even before its text began, its paratexts offered signs that it would occupy a contestable space between genres. Its front cover featured a photograph of Feinberg – a stylised, airbrushed image that recalled 1980s digital portraiture, simultaneously representing and de-familiarising the author.

The back cover's paratextual material opens with 'Woman or man? That's the question that rages like a storm around Jess Goldberg, clouding her life and her identity', before describing Goldberg's 'growing up differently gendered' and working class in the 1950s, coming out as a butch in the following 'prefeminist' decade, and 'deciding to pass as a man in order to survive' in the 1970s. The author's own biography is briefly summarised below a more naturalistic photo, stating that 'Leslie Feinberg came of age as a young butch in Buffalo, New York, before the Stonewall Rebellion.' A disclaimer states: 'This is a work of fiction. Any similarity between characters and people, dead or alive, is a coincidence' (Feinberg 2). The acknowledgements, however, explicitly linked the text back to Feinberg's own experiences: 'There were times, surrounded by bashers, when I thought I would not live long enough to explain my own life. There were moments when I feared I would not be allowed to live long enough to finish writing this book. But I have!' (4).

The novel itself is a *Bildungsroman*, explaining, in the first person, how Jess Goldberg comes to define as a stone butch (a masculine lesbian who does not like to be touched) through an engagement with the queer urban subcultures of New York before the Stonewall riots, and an understanding of how those subcultures intersect with a hostile wider world. In it, Goldberg is raped; beaten during police raids on the bars where the butches meet with femme lesbians, drag queens and other outsiders; and set up to be injured by a factory floor manager, who distrusts not just Goldberg's gender identity but also hir Jewishness and trade union activity. (Feinberg wrote about pronouns for people who did not fit into 'him' or 'her', 'he' or 'she', or 'he-she', used as an insult and reclaimed in *Stone Butch Blues*. I use 'ze' and 'hir', as Feinberg preferred.)

Although the narrative begins with 'I didn't want to be different' (which, after all, *is* often true for gender-variant individuals) and the pain of not fitting into the categories of 'boy' or 'girl', Goldberg does not move from the fraught 'he-she' position towards maleness until halfway through the novel. At this point, it feels more like the older transsexual memoirs, in that it explores the challenges of shifting from one gender position to another rather than the struggles of occupying a space between or outside them (Feinberg 13). The scenes at the beginning of chapter 15, in particular, feel like classic transition memoir fare: the mirror scene, where Goldberg first notices the effects of testosterone and the joy at realising that breast reduction surgery is almost affordable (Goldberg never seriously contemplates phalloplasty, perhaps as it was prohibitively expensive and not always effective during the post-war period); the strange thrill of being called 'sir' on daring to 'enter men's turf' at the barber; and getting through 'the most important test of all: the men's room' (Feinberg 171–72). In many other accounts, such 'success' would be an important step towards resolution, but Goldberg turns away, and relates that opening some avenues closes others, as getting male identification in order to drive (or work) may prove impossible (Feinberg 175).

The first scene of Goldberg buying men's clothes anticipates this refusal to blend in, being conducted not with a view to avoiding attention, but attracting it, as 'three powerful queens in full drag' come to help Jess choose a suit to wear while compering a show at the Malibou club (Feinberg 58). Later, in trying to 'pass' within straight society, Goldberg worries about selling out those who couldn't, and the butches' admiration for this ability only amplifies hir guilt. As uncomfortable with being seen as male as female, Goldberg stops taking testosterone, and feels ambivalent about hir position between traditional roles, mainly because it leads others to police hir gender:

> My hips strained the seams of men's pants. My beard grew wispy and fine from electrolysis. My face looked softer. Once my voice was hormone-lowered, however, it stayed there. And my chest was still flat. My body was blending gender characteristics, and I wasn't the only one who noticed. (Feinberg 224)

Two-thirds of the way through the story – rather than at its climax, like Jan Morris – Goldberg sets up the 'epiphany' of sex reassignment surgery, only to reject it, and shift the focus to the social: '"I've seen about it on TV. I don't feel like a man trapped in a woman's body. I just feel trapped."' (Feinberg 158–59)

In his ground-breaking *Second Skins: The Body Narratives of Transsexuality*, the first book-length study of such narratives from a trans perspective, Jay Prosser devotes a whole chapter to 'Transgender and Trans-Genre' in *Stone Butch Blues*. He writes that it 'does not abandon but reconfigures the conventions of transsexual narratives' by refusing the closure of 'fully becoming the other sex' despite hormone use and some surgery, as

Goldberg chooses 'an incoherently sexed body in an uneasy borderland between man and woman' (Prosser 178). *Stone Butch Blues* complicates the old transition narrative and subverted one of its biggest clichés – the bodily 'homecoming' that follows sex reassignment surgery – as the comfort that came after Goldberg's top surgery dissipated with the realisation that 'passing' as male did not feel like a route home. Through a close reading of the work alongside the 'extra-text' that surrounded and shaped its reception, Prosser argues that Feinberg's evasion of the 'homecoming' trope relies upon the ambiguity over whether *Stone Butch Blues* should be taken as fiction or memoir.

How deliberate this ambiguity was is up for debate: Feinberg certainly found it liberating, as ze confirmed in several contradictory statements. In email correspondence with Prosser in 1996, Feinberg insisted that although *Stone Butch Blues* 'drew on my knowledge of what industries and avenues were open' to a trans person, the 'emotional and situational path, transgender path choices and consciousness of [Jess Goldberg]' were fiction (Prosser 191). However, in an earlier interview for an FtM (female-to-male) newsletter, Feinberg said:

> I felt, by telling it autobiographically, that I would pull back in a lot of places … as transgendered people, that we're always being told who we are, either physically or emotionally – strip or be stripped, you know? … I feel we've each found our own boundaries of dignity which we will not go beyond; that we deserve. I really felt that by fictionalizing the story, that I would be able to tell more of the truth; be more brutally honest than I would if I were telling my own story. (Prosser 192)

Prosser focuses on Goldberg's ambivalence about 'passing' as a man, and *Stone Butch Blues*' reticence about 'passing' as a novel, and its radical refusal of closure; he notes Feinberg's distinction between 'truth' and 'facts', and how hir narrative draws parallels between the police that strip Goldberg in queer bars – as a pretext to more humiliation and assault, and a violent act in itself – and the autobiographical pact that demands authors to 'reveal' themselves through disclosure of verifiable details. Considering this, Feinberg's use of even a 'bad' pseudonym ('Jess' and 'Leslie' are ambivalently gendered and 'Goldberg' resembles 'Feinberg', ethnically and phonemically) is enough to buy hir out, and keep *Stone Butch Blues* in that trans-genre space: its events do not need to have happened to Feinberg, but to have plausibly happened to members of the trans/lesbian subculture of which Goldberg is a part (Prosser 196–97).

By taking *Stone Butch Blues* into a genre position that is first clarified as 'A Novel' and then confused, Feinberg managed to escape not just the conventions of the transsexual memoir, but also many of its pressures. Whilst not primarily an imperative aimed at trans people, like hir *Transgender Liberation* pamphlet (1992), it did not convey a sense of striving to justify itself to outsiders like most of the texts after Hedy Jo Star's *I Changed My Sex* (1953), named by Stone as 'the first fully autobiographical' transsexual account (224). Feinberg could not avoid *Stone Butch Blues* being read through the prism of hir own life, however, and the publisher's framing did not help (although one would be surprised if ze did not have much input/influence over this), although its contemporary critics were largely from the LGBT community, and so, like Prosser, appreciated the author's reasons for its being 'trans-genre'.

This relationship between queer narratives, the 'extra-text' and authenticity was pushed to the limit at the turn of the millennium, by the emergence of J.T. LeRoy's novels *Sarah*

(1999) and *The Heart is Deceitful Above All Things* (2001), his subsequent adoption into the celebrity wing of American counter-culture and the 2005 revelation that LeRoy was not an 18-year-old, HIV-positive former hustler with gender dysphoria from California, as was claimed six years earlier, but a New York-based thirty-something mother named Laura Albert who had persuaded her sister-in-law (Savannah Knoop) to be 'LeRoy' in public. Amidst the anger of embarrassed people who had helped what they saw as a troubled youth and called LeRoy 'a hoax', Warren St John's *New York Times* exposé (9 January 2006) raised a fascinating issue: 'It is unclear what effect the unmasking of Ms. Knoop will have on JT Leroy's readers, who are now faced with the question of whether they have been responding to the books published under that name, or to the story behind them.' ('The Unmasking of JT Leroy')

It seemed that readers were not responding *either* to the text *or* to the author, but the intersection between them: it was not just that the novels were good, but also that they apparently came from someone so young and disadvantaged, and drew on that past, incorporating numerous signifiers that seemed chosen to appeal to those enthralled US counter-culture – homelessness and hustling in small-town America, gender ambiguity and sexual exploration, with a very 1990s glamorisation of being a 'fucked up' teenager. There was a cruel irony in the contrast between Feinberg's aim of liberating trans writers from the memoirist's obligation to stick to the facts and some trans readers' desire for an authentic experience behind LeRoy's texts, and disappointment in finding out that someone who apparently took a trans voice beyond specialist LGBT publishers (such as Firebrand) and into the literary mainstream turned out not to be a trans woman, even if Albert's defence that the books had always been sold as 'fiction' was accurate. (One answerphone message recorded by Albert and featured in Jeff Feuerzeig's documentary *Author: The JT LeRoy Story* (2016) said 'the transgender community is going to want to fucking lynch you.')

*

I had somehow missed the LeRoy affair at the time, even though it played out as I was trying to establish myself as a trans-identified writer of cultural criticism and fiction. Trans issues were not the only thing I wrote about by any means – most of my published journalism and (largely) unpublished short fiction covered literature, film or art – but it was important to me to write about gender identity. In my youth, I had lacked a vocabulary to understand my gender dysphoria, and become frustrated with the ignorance or hostility with which trans people and politics were covered in newspapers and magazines, film and television. Bornstein, Feinberg and Stone had given me a language – not just specific terminology, but an entire paradigm through which I could better think about my body and how it interacted with the wider world – but a decade or more since their texts were first published, I did not see their perspectives anywhere within the mainstream media.

What I *did* find in the mid-2000s, in the apparently 'progressive' *Guardian*, were several pieces reiterating Janice Raymond's take on transsexual people, watered down to fit within the 'respectable' limits of liberal-left discourse. Julie Bindel's 'Gender benders, beware' article of January 2004 became a touchstone for trans activists who organised online: Bindel's piece, published in print and on the *Guardian* website, began by supporting Vancouver Rape Relief's decision not to allow a transsexual woman to train as a counsellor for female rape survivors, and moved into a broadside against trans people in general, concluding that: 'I don't have a problem with men disposing of their genitals, but it does not make them women, in the same way that shoving a bit of vacuum hose down your 501s does not make you a man.'

By the summer of 2009, with the help of the first wave of transgender theorists and successors such as Julia Serano, I had come to define as 'transsexual', and sought a referral to the Gender Identity Clinic in West London in the hope of getting hormones and sex reassignment surgery. A friend suggested that I pitch a blog to the *Guardian* site, documenting the physical and social experiences of transition in real time. The paper's coverage of trans issues had become mired in arguments about 'freedom of speech' ever since Bindel's article had prompted over two hundred complaints and demonstrations at LGBT events where she appeared – an organised opposition that columnists such as Beatrix Campbell labelled 'censorship' ('Censoring Julie Bindel'). This, I felt, should not be taken at face value, especially as the characterisation of trans issues as too 'academic' for mainstream audiences (notably in Julie Burchill's subsequently deleted *Observer* article, 'Transsexuals should cut it out' of January 2013) served to exclude trans people from a political discourse in which we were attacked from all sides (Pugh).

A rolling blog, I thought, might change the nature of the *Guardian*'s trans discussions, and allow me to challenge the stereotypes and assumptions on which those hostile feminists relied, without having to argue on the terms that they set (and so getting lumped in with 'censorious' trans activists who did so). In a journalistic discourse that felt like it was 20 or 30 years behind activist and academic ones, borrowing a strategy from Carol Riddell and Sandy Stone seemed like an effective tactic. The limitless space available online (with the low fees for blogs – usually £90 per post – meaning it represented little financial risk for the *Guardian*) meant I could follow Stone's imperative to 'mix genres' by bringing in trans theory and cultural history to get beyond the 'long-suffering romantic stuff' that Bornstein had lamented in the 1990s, while still using a framework that the *Guardian* editors would recognise and, thus, permit. (It was their choice, not mine, to entitle it *A Transgender Journey*.)

In the process, I hoped to demonstrate to print editors and broadcast commissioners that wider audiences would be interested in (and understand) trans people's writing about our personal and political concerns if we were allowed to present it to them. The space I had targeted was vital: there were many transition blogs by 2009, but none in a publication with the *Guardian*'s reach – I knew from my various mid-2000s office jobs that many people who would not seek out trans theory might click on (or share) my articles, and be challenged by them, thus eroding the (largely unspoken) borders between cis and trans audiences. This was helped further by having an open comments section, which allowed a community to grow around the series and show that support for, and hostility to, trans people cut across lines of class, sexuality and gender. Importantly, it provided a forum for other trans people to outline their experiences, reinforce (or contradict) my arguments and voice opinions about the newspaper's past coverage, without needing to name specific writers. This, I hoped, might offer a transformative perspective for cis readers, encouraging them to rethink the terms on which they read about our lives – something that I tried to raise directly when it felt appropriate.

After getting the series commissioned and finding a large audience with the first three pieces, I was allowed to write as many entries as I required to complete my narratives and get my points across. Eventually, I wrote 30 articles between 2010 and 2012, which allowed me to unpack some long-established assumptions. In the 16th article, long after I had established myself as a character and explained the basics of the reassignment process, I could reach more complex conclusions about ideas such as being 'trapped in the wrong body', which had long since become a cliché in descriptions of transsexuality. Instead, I could state that I had never quite felt like this; it was more the case that I could only function 'if I re-launched the symbiotic relationship between my body and mind from a starting position that felt right.' (Jacques)

After I published the 3000-word article about my sex reassignment surgery – having little choice but to replicate the climactic structure of previous narratives – an agent approached me about turning the column into a book. Ever since the series had begun, people had asked if I intended to do so: I always said no, because the important thing had been the blog format, as a Trojan horse to take trans history, theory and politics to a wider audience, with the rolling publication meaning that the surgical peak was not pre-determined. There were numerous books telling transition stories, and I thought the entire genre was dated: I struggled to see what re-writing *A Transgender Journey* might achieve. After conversations with trans friends, I felt that writing about something else (as I did in my *New Statesman* blog that I secured through the *Guardian* series, in which I wrote mostly about the arts) would be more politically useful, showing young trans people that they did not have to write only about their own lives.

Having considerably raised my profile, I found that agents and publishers were coming to me for a book, but they declined my suggestions for other projects (such as a history of trans people in Britain, or a collection of short stories that included several about trans individuals) in favour of a memoir based on the *Guardian* blog. If I wanted to write fiction, plays or screenplays – as had been my original goal – then I would have to do a memoir first.

Working within this genre, I would be bound by Lejeune's autobiographical pact, needing not just to convey 'the truth' of my experiences but also to stick to the facts. My efforts to avoid the obligation to 'strip or be stripped' would have to be made via its structure, direct address to readers, and the inclusion of material that unpicked some of the clichés and conventions of trans memoirs and other media coverage, rather than by obfuscating the character of its protagonist like 'auto-fiction' authors such as Chris Kraus or Sheila Heti (whose novels had central characters who shared their names, with narratives that invited readers to guess what had happened and what was invented). Even writing 'A Novel' like *Stone Butch Blues* did not feel plausible: with the *Guardian* series behind it, I felt that any first-person narrative would be read as autobiography, negating the freedom that Feinberg had mentioned to Prosser.

Despite these limitations, I accepted an invitation from Verso Books to write *Trans* in July 2013. With the contract signed, I had to work out how to make the book sufficiently different to the blog to make it worthwhile – and how to avoid some of the traps of the transition memoir. I decided to cross-cut between chapters about my life and interludes about representations of trans people in the media to avoid the problem of memoir not placing individual experiences within a wider context, and included a conversation with Sheila Heti as an Epilogue, where we discussed the possibilities and frustrations of documenting trans experiences in mainstream spaces, and talked explicitly about how transition 'didn't feel' like a 'mythical hero's journey' but 'a bunch of hoops to jump through while working in boring jobs' (294). I got around the pitfall of building up to surgery as the climactic point by using a facsimile of my *Guardian* account as the opening chapter, in part to establish that my book would simultaneously be about my transition, and my relationship with the media, and the psychological effects of those two things becoming intertwined. As *A Transgender Journey* changed my life – launching me into different social circles in a different city, as I moved from Brighton to London to pursue my writing career – it was not difficult to bring this subject matter into *Trans: A Memoir*.

Indeed, my favourite passage of the book came near the end, when I reflected on my visit to a glamorous literary party at a prestigious London venue:

> I'd felt like an outsider all evening, thinking about how comment journalism wasn't for me, with its strident with-us-or-against-us arguments, and how tiresome I found it when prominent columnists claimed not to be part of 'the Establishment'. Even if my perspective was unusual within those circles, I was now definitely part of them. (257)

In conclusion, I noted that 'If you articulate an outsider critique well enough, you stop being one' (302), referring to the way in which I was invited to meet many of the journalists whose views I had opposed, and who expressed an interest in my work. I began to rely on them for access to spaces where I might counter their opinions – and felt that I could not single anyone out, or be too vociferous in my criticisms of them, or of the media in general.

The book was published in September 2015 with the subtitle of 'A Memoir' – as a conscious nod to Feinberg, as well as a rejection of 'more commercial' ones suggested by Verso, such as 'finding the real me'. (The cover echoed *Stone Butch Blues*, as I declined my editor's suggestion that I use a photograph and asked that writer/illustrator Joanna Walsh draw me; Australian trans author Tom Cho's semi-autobiographical short fiction collection *Look Who's Morphing* [2009] employed a similarly stylised image.) But for all the ways in which I tried to resist the formal limitations of memoir, I still had the fundamental problem that Feinberg described two decades earlier. Every chapter I sent to my editor came back with a request to be 'more personal' – as had happened with the *Guardian* – but I didn't feel comfortable in going deeper into my mental health issues, my sexuality or my relationship with clothes when transphobic people often attacked us as fetishists, lunatics or superficial, and trans theorists expressed concerns that autobiographical writing might play into their hands.

In *Trans: A Memoir*, I had to react against the idea that memoir was for cisgender outsiders whilst theory was for trans people by attempting to combine personal experiences and political reflections, aiming for trans readers who were still working out their identities. Having completed it, I felt compelled to return to fiction, thinking that it might enable me to entirely avoid that dichotomy. Certainly, its flexibility allowed me to get around framing transness as a 'problem' (as trans activist, poet and publisher Cat Fitzpatrick put it in her acute assessment of my book and the structural limitations of memoir for Lambda Literary) by focusing on a smaller section of someone's life than the long self-realisation of transition, or putting characters in situations where their genders complicated their relationships or goals.

I preferred short fiction as a means of avoiding a centralised narrative – and, of course, of detaching the writing from my own life, about which I felt I had nothing more to say, at least not in an autobiographical format. I did not make any distinctions about whether the text was for 'cis' or 'trans' readers: I hoped that my extensive first-person documentation would lend credibility, but that the authenticity of my characters' experiences would come through in my texts. I did not limit myself to writing about trans people, but the tension I'd had in journalism remained – of not wanting to feel restricted or typecast, but also feeling that covering these issues was important. So, I published fiction about various subjects, not worrying too much about whether I brought a trans perspective to them, but returned to my plan for a body of trans short stories, partly in hope of using a different form to further change the terms of discussion about trans lives.

Before it came out, an editor at Catapult asked for a piece that would tie in with it. Initially, she requested an essay that would interweave my life experiences with a reading of post-war *avant-garde* author Ann Quin. Via email, I suggested a 'semi-fictional response', set in mid-2000s Brighton, that built on Quin's themes (masculinity and gender, the disappearance of Brighton's 'old' end-of-the-pier entertainment, and the shabbiness of the town). This would let me further explore the time and place evoked in my memoir, but use material that did not fit, or that I was reluctant to include for fear that it would be attached to me, or the wider community.

One problem I'd had with writing a memoir was that throughout my life I'd tried to explore and express my gender identity with as little drama as possible. In the resultant story, *Weekend in Brighton*, I could imagine how situations where I had avoided conflict might have turned out differently, and felt able to explore a trans woman's sexuality 'as the disruptive, excessive reality and experience it mostly is', as Jacqueline Rose put it in a *London Review of Books* essay prompted by my book, entitled 'Who do you think you are?', where she noted that the textual obligation to hide ambivalence about 'passing' meant that most memoirs (including mine) made little space for sex (12). To discuss sex, especially any pre-transitional encounters, would require me to talk extensively about my body, and about my genitalia, running the risk that readers would unambiguously read me as male, undermining the identity that I had carefully set up in my text, and in my life. I thought about combining the material of memoir and the theoretical imperative set by Stone (that Rose also mentioned) and drew my experiences of stultifying post-graduate jobs, having little money, exploring my gender and sexuality without anyone to guide me, meeting others who had been in similar isolation and trying not to let our internalised transphobia damage our relationships, into a single story.

One discovery was that my central character did not need to be labelled, or self-label, in terms of gender – I could introduce the underlying issue with 'today, as most days, he *was* Patrick, not Trish' but avoid words such as 'cross-dresser', 'transvestite' or 'transsexual' (Jacques). This moved the focus away from those identities and towards sexuality, with a sense that the 'butch' and 'femme' venues that had survived the AIDS epidemic and the assimilationist gay and lesbian politics that followed were not welcoming towards gender-variant people; I found at the time that fetish clubs were a more open environment, and so too does Patrick/Trish.

In a scene in a shop that sells clothes and accessories to male-to-female people, protagonist Patrick/Trish (whose pronouns shift with gender presentation) bonds with Brian/Bree – an older person who also has a female persona. This connects them before they meet at the Harlequin, a trans-friendly club that often had drag performances, which I talk about in my memoir (Jacques, 48–51, 73–74, 95–97). In the first draft of *Trans: A Memoir*, I also documented a visit to its S/M night, but my editor told me to remove it;

I drew on the unpublished journal material that I had used for those expurgated passages into the climactic scenes of *Weekend in Brighton* where Trish and Bree find that they enjoy spanking each other in the club, but Trish goes to Bree's hotel to find that she cannot relax into similar interaction for fear of attracting attention. Ultimately, their connection does not survive as Trish retreats from her own desires, but the story could address ways in which sexuality can act as a motor for personal developments when the memoir could not, as its focus remained primarily on my transition.

When Audible turned *Trans* into an audio book, they asked me to record extra material: I chose *Weekend in Brighton*, further positioning it as fiction that was close to memoir, and blurring the genre lines in Feinberg's spirit. This meant the story could reach a wider audience – as trans memoir, thanks to the workings of the market and the prejudices of editors and publishers, remains far more prominent than fiction. It also allowed me to bring more explicit sexual content into the memoir's orbit, but without intruding upon its structure, and indeed, it is structure that is crucial: it may be that the traditional memoir's purpose, in 'normalising' transition for a cisgender audience, is now less useful, socially, than finding new ways to tell stories that can express an infinite variety of gender positions, currently being explored via blogs and social networks.

Disclosure statement

No potential conflict of interest was reported by the author.

References

Author: The JT LeRoy Story. Directed by Jeff Feuerzeig. Vice Films, 2016.

Bindel, Julie. "Gender Benders, Beware." *The Guardian* 31 Jan. 2004. Web. 19 Aug. 2016. <www.theguardian.com/world/2004/jan/31/gender.weekend7>.

Bornstein, Kate. *Gender Outlaw: On Men, Women, and the Rest of Us*. New York: Vintage, 1995.

Bornstein, Kate. *A Queer and Pleasant Danger*. Boston: Beacon, 2012.

Campbell, Beatrix. "Censoring Julie Bindel." *The Guardian* 31 Jan. 2010. Web. 19 Aug. 2016. <www.theguardian.com/commentisfree/2010/jan/31/julie-bindel-transgender-nus>.

Feinberg, Leslie. *Stone Butch Blues: A Novel*. New York: Firebrand, 1993.

Fitzpatrick, Cat. "'Trans' by Juliet Jacques." *Lambda Literary* 3 Nov. 2015. Web. 8 Aug. 2016. <www.lambdaliterary.org/reviews/11/03/trans-by-juliet-jacques>.

Harrod, Horatia. "The Tragic True Story Behind *The Danish Girl*." *The Daily Telegraph* 28 Feb. 2016. Web. 27 July 2016. <www.telegraph.co.uk/films/2016/04/14/the-tragic-true-story-behind-the-danish-girl>.

Hirschfeld, Magnus. "Selections from The Transvestites: The Erotic Drive to Cross-Dress." *The Transgender Studies Reader*. Eds Susan Stryker and Stephen Whittle. New York: Routledge, 2006. 28–39.

Jacques, Juliet. “Finally, Slowly, My Body was Catching Up with My Mind.” *The Guardian* 12 Jan. 2011. Web. 19 Aug. 2016. www.theguardian.com/lifeandstyle/2011/jan/12/transgender-health-and-wellbeing.

Jacques, Juliet. *Trans: A Memoir*. London/New York: Verso, 2015.

Jacques, Juliet. “Weekend in Brighton.” *Catapult* 20 Oct. 2015. Web. 24 Aug. 2016. <catapult.co/stories/weekend-in-brighton>.

Marcus, Laura. *Auto/biographical Discourses: Criticism, Theory, Practice*. Manchester/New York: Manchester University Press, 1994.

Morris, Jan. “Conundrum.” *Sexual Metamorphosis: An Anthology of Transsexual Memoirs*. Ed. Jonathan Ames. London: Vintage, 2005. 77–98.

Prosser, Jay. *Second Skins: The Body Narratives of Transsexuality*. Chichester/New York: Columbia University Press, 1998.

Pugh, Andrew. “Mulholland Pulls Burchill’s ‘Disgusting’ Transsexuals Rant: ‘We Got It Wrong’.” *Press Gazette* 15 Jan. 2013. Web. 19 Aug. 2016. <www.pressgazette.co.uk/mulholland-pulls-burchills-disgusting-transsexuals-rant-we-got-it-wrong>.

Queen, Carol, and Lawrence Schimel, eds. *PoMoSexuals: Challenging Assumptions about Gender and Sexuality*. San Francisco: Cleis, 1997.

Raymond, Janice G. “Sappho by Surgery.” *The Transgender Studies Reader*. Eds Susan Stryker and Stephen Whittle. New York: Routledge, 2006. 131–143.

Riddell, Carol. “Divided Sisterhood: A Critical Review of Janice Raymond’s *The Transsexual Empire*.” *The Transgender Studies Reader*. Eds Susan Stryker and Stephen Whittle. New York: Routledge, 2006. 144–158.

Rose, Jacqueline. “Who Do You Think You Are?” *London Review of Books* 5 May 2016. Web. 27 Feb. 2017. <www.lrb.co.uk/v38/n09/jacqueline-rose/who-do-you-think-you-are>.

St. John, Warren. “The Unmasking of JT Leroy: In Public, He’s a She.” *New York Times* 9 Jan. 2006. Web. 3 Aug. 2016. <www.nytimes.com/2006/01/09/books/09book.html>.

Stone, Sandy. “The Empire Strikes Back: A Posttranssexual Manifesto.” *The Transgender Studies Reader*. Eds Susan Stryker and Stephen Whittle. New York: Routledge, 2006. 221–235.

Confessional Poetry and the Materialisation of an Autobiographical Self

Maria Takolander

ABSTRACT
Little has been said about what happens when the writing subject cobbles together a self from the material of language in an always-unpredictable creative experiment. This is undoubtedly in part because life-writing scholars are not conventionally life-writing practitioners. In this article, informed by my experience writing confessional poetry and by socio-material scholars of creativity such as Vlad-Petre Glăveanu, I explore how the technologies of poetic language – bearing affordances and constraints, embodying social and cultural histories, and therefore exerting their own intentionality – are exercised in the creative act, giving rise to a phenomenological alienation of the autobiographical subject that is often explained in terms of unconscious forces or 'madness'. I begin by discrediting myths linking creativity and mental illness in order to destabilise a traditional view of confessional poetry as generated by individual pathology, before going on to theorise the phenomenological lacuna in authorial subjectivity that such myths represent in relation to the potent technologies of poiesis. Ultimately, this article argues that confessional poetry arises from a distributed socio-material practice in which the agency of the author is negotiated through the agency of writing materials in the creation of autobiographical artefacts, thus exposing another limit in the field of life writing.

Life writing has long been theorised in terms of its limits. Scholars have discussed the 'representation fallacy' of the autobiographical text and the 'ontological fallacy' of the autobiographical subject (Brockmeier and Harre 48), that complex and chimerical being given shape and coherence by narrative conventions of unity and verisimilitude. We have heard about how memory, particularly when it involves trauma, is unreliable and temporally constructed; how the self is also always not itself, being based in intersubjective relations; how the subject (in an Althusserian or Foucauldian sense) is an effect of power relations in society. As Sidonie Smith and Julia Watson put it, the self has become 'the source, authenticator and *destabiliser* of autobiographical acts' (22, emphasis added). Taking inspiration from Bruno LaTour's actor–network theory – acknowledging objects and technologies as well as humans as actors in constructing worlds – and theories of distributed cognition more generally, this article will add to the destabilisation of the autobiographical subject by attending to the agency of the materials involved in the creative act of life writing.

If we have consistently heard that 'it is not possible to separate the "life" from its textualisation … its writing and its reading' (Gill "Your story" 72), little has been said about what actually happens when the writing subject cobbles together a self from the material of language in an always-unpredictable experiment with the proverbial pen and paper or, moreover, keyboard and screen. This is undoubtedly in part because life-writing scholars are not conventionally life-writing practitioners. However, emboldened by my experience writing confessional poetry and by scholars theorising creativity as a socio-material practice, such as Vlad-Petre Glăveanu, I explore how the technologies of literary writing – bearing affordances and constraints, embodying powerful social and cultural histories, and therefore exerting their own intentionality – give rise to a phenomenological alienation of the writing subject, exposing yet another limit in the field of life writing.

This article focuses on confessional poetry, informed by my experience of writing a sequence of confessional poems for a new collection (with completed poems published in various national and international forums[1]). Poetry is, of course, already a limit case in life writing, even though critical attention to the confessional poems of Sylvia Plath and her contemporaries has been crucial to the development of the field – and, indeed, to an awareness of its limits, particularly with regard to how autobiography, as a literary practice, 'constitutes, as much as it transforms, the reality to which it is presumed to refer' (Rose 108). As Melanie Waters and Jo Gill write in their introduction to *Poetry and Autobiography*, 'poetry, because of its peculiar, self-conscious, often explicit negotiation with the dynamics and parameters of language, is precisely the place where some of the key issues, concerns and debates in autobiography studies might be most keenly and usefully played out' (2). Because of its foregrounding of 'the phenomenal *materiality* of language' (Schleifer 85) – those cadences and rhymes informed by orality, often rendered spatially on the page – poetry is certainly a useful genre for understanding how the agency of writing technologies powerfully impact on the agency of the subject engaged in the event of life writing.

In *On Knowing: Essays for the Left Hand*, Jerome Bruner describes how, during the creative act, a writer embraces the '[*f*]*reedom to be dominated by the object*' (25, original emphasis). Putting aside a traditional view of creative practice as enabled by a sovereign moment of inspiration, Bruner explains:

> You begin to write a poem. Before long it, the poem, begins to develop metrical, stanzaic, symbolic requirements. You, as the writer of the poem, are serving it … It is at this point that we get our creative second wind, at the point when the object takes over. (25)

Poiesis is configured here as an emergent practice resulting from engaging with a writing techne in ways that resonate with my experience and in ways that are important for this article's investigation into life writing. The poem's rhythms and rhymes, the images and metaphors that emerge through a sensitivity to the developing form and content of the poem, the kinds of events and emotions that these things recall, domain-specific knowledge about other poems (both explicit and implicit); all contribute to give rise to the textual materialisation of a self over which one does not have complete control. The poem and the self are exposed through this process as 'open systems' (Glăveanu "Principles" 158) in mutual interaction; the sharing of agency between the writing technology and the writing subject is what produces the creative work. It is in this way that a confessional poem (and, indeed, any act of life writing) – as well as the selves constructed therein – must always be understood as aesthetic acts. It is also through investigation of this creative process that the often-noted 'artful manipulation' of confessional poetry (Gregory 35) – which suggests a greater agency at work on behalf of the writer than is necessarily present – can be newly understood.[2]

Indeed, an appreciation of the agency of writing technologies is particularly pertinent to confessional poetry, given how the creative identities of its key practitioners – Plath, Robert Lowell, John Berryman, Anne Sexton – loom so large in the popular and scholarly imagination. Interpreters of Plath's poetry, for example, as Tracy Brain argues, can sometimes be 'blinded by biography' and, notably, seem especially focused on authenticating representations of 'emotional extremity' (13). Indeed, it is one of the arguments of this article that the attribution of the work of the confessional poets to intense emotion or mental illness significantly contributes to the powerful aura of their authorship, informed by a time-honoured ideology of 'auto-intoxicated' creativity (Moffitt 20) in ways that further negates the agency of writing technologies. It is for this reason that my article begins by contesting the expressive view of poiesis bound up with the myth of the mad poet and of the *furor poeticus* – while noting how such a characterisation of creativity usefully gestures towards a lacuna in authorial agency, even if its ascription to psychopathology or, more generically, the 'unconscious' is ultimately limiting. Following this, I outline a socio-material theory of poiesis that emphasises the agency of the aesthetic medium of poetry. The point is not to deny the relevance of the autobiographical impetus of confessional poetry – or that writers of confessional poems have suffered mental illness – but, rather, to acknowledge the sharing of agency between the autobiographical subject and her writing technologies in the scene of life writing. This is, as Glăveanu puts it, an argument against '*individualism, not the individual*' (*Distributed Creativity* 9, original emphasis), and for an understanding of creativity as 'distributed action' (23) rather than as a spontaneous output of 'the "box" of the individual mind' (2).

Contesting madness as methodology

As Frederick Burwick documents in *Poetic Madness and the Romantic Imagination*, the figure of the mad poet and the methodology of the *furor poeticus* has a 'venerable' cultural history (3). The first critics of Western literature, Plato and Aristotle, theorised poiesis as a form of madness, according to which creativity was explained by divine and frenzied inspiration. In the modern world, 'the century-old notion of the *furor poeticus*' was famously valorised by the Romantic poets 'as a revolutionary and liberating madness that could free the imagination' (Burwick 2) and – paradoxically, given the sense of impersonality implicit in the original myth – give rise to a true individualism. The Romantic ideology of individualism, of course, also gave rise to modern forms of life writing – and pertinently, as Laurence Lerner argues, to the idea of 'the poet as biographical individual', placing 'emphasis on personal immediacy, on the poet as a man speaking to men, and on the true voice of feeling', and encapsulating an ideology of poiesis which 'has been with us ever since *Lyrical Ballads*' (48). Centuries later, the confessional poets reasserted these links between creativity, individuality and feeling (or 'madness'), reacting against a period of repressive social conformity – the US Cold War era of the 1950s and 60s –

and against the impersonal aesthetic of High Modernism and New Criticism. In fact, as Lerner reminds us, A.A. Alvarez, one of the early critics of confessional poetry, conceptualised Lowell's work as '"an extension of the Romantic agony into modern, analytic terms"' (quoted in Lerner 47).[3]

As Albert Rothenberg observes in *Creativity and Madness: New Findings and Old Stereotypes*, psychoanalytic theories of creativity have become thoroughly 'ingrained' in thinking about the creative process (48) – and, certainly, a mystical 'belief in the unconscious roots of creativity' (48), or a clinical articulation of this general principle within the discipline of psychology, are central to how confessional poetry has been received. This makes sense given that the fields of psychoanalysis and psychology had a significant impact on the confessional poets, who belonged to a middle-class 'coterie' (Lerner 66) of American poets for whom Freudian psychoanalysis, breakdowns, psychotropic medications and institutionalisation formed shared life and artistic ground. The confessional poets were, in other words, self-consciously and, for some, tactically 'mad'.[4] Certainly, the apotheosis of the confessional poets as great artists cannot be isolated from the ways in which their art confirmed a culturally longstanding and powerful association between creativity (particularly poiesis) and 'madness'.[5]

It is an association that continues to have considerable cultural traction and, moreover, to be taken seriously. In *Creativity and Mental Illness*, Simon Kyaga outlines 98 contemporary studies in the discipline of psychology that explore the connection between creativity and psychopathology (defined in relation to the Diagnostic and Statistical Manual of Mental Orders) (107). In most of these studies, the confessional poets are key to the evidence. The psychologist Kay Redfield Jamison's 2017 biography of Lowell, for example, doubles as a case study detailing the links between creativity and psychopathology. Redfield Jamison presents Lowell's bipolar disorder as a 'determining force' in relation to his writing (283), literally conceptualising poiesis in terms of bipolar symptoms: '[d]epression is a ruminative, highly self-critical state, ideal in its way for revising work generated in a spewing, generative, less self-censoring manic state. Depression prunes and edits … Depression also corrects' (302–3). For Redfield Jamison, 'any attempt to understand Lowell's work must necessarily be "more seismographic than aesthetic"' (293). This is ironic, given Redfield Jamison's penchant for aestheticising Lowell's mental illness – a practice for which we can find a context and a corrective in Susan Sontag's *Illness and Metaphor*, a study that in part traces the nineteenth-century Romantic aestheticisation of the sufferer of tuberculosis as a 'hectic, reckless creature of passionate extremes, someone too sensitive to bear the horrors of the vulgar, everyday world' (36). The sufferer of TB was likewise represented as a natural poet, with John Keats and Percy Bysshe Shelley providing famous examples. Sontag is explicit in connecting the Romantic aestheticisation of TB with the contemporary aestheticisation of mental illness: 'In the twentieth century, the repellent, harrowing disease that is made the index of a superior sensitivity, the vehicle of "spiritual" feelings and "critical" discontent, is insanity' (36). What motivates both myths, for Sontag, is 'that distinctively modern activity, promoting the self as an image' (29) – one essential to life writing.

Notwithstanding Sontag's work, however, the explanation of confessional poetry as determined in some sense by pathology remains consistent even in the humanities, though there is a marked preference for psychoanalytic theory over the terminology of the DSM-5. In Rose Lucas's essay on the poetry of Anne Sexton, for example, she argues: 'Sexton's art rises directly out of the concomitantly oppressive and potentially cathartic experiences of depression, or melancholia, and the extreme psychotic or manic breaks for which she was repeatedly hospitalised and treated' (46–7).

There are various reasons to be suspicious of scholarship strongly correlating creativity with 'madness', as Sontag's research suggests and as I have argued at length elsewhere ('"A Dark/Inscrutable Workmanship"' and 'From the "Mad" Poet').[6] When it comes to confessional poetry specifically, one might start by pointing out how the 'representation fallacy' and 'ontological fallacy' (Brockmeier and Harre 48) noted earlier often seem to go unscrutinised. This is despite the fact that, as numerous critics have noted, there is clearly much more going on than organic self-expression in these carefully crafted poems. For one, the confessional poet is self-consciously working within a genre – autobiography and, more specifically, confessional poetry – and for an audience. Accordingly, the confessional poets 'put into play a reality trope ... to convince readers that their work deserves attention' (Gregory 36), and abide by the requirement of '*exceptionality*' (Brockmeier and Carbaugh 29, original emphasis) in constructing a dramatic identity for performative effect – as theatrical readings by Plath, Berryman, Lowell and Sexton (all readily available on the internet) attest. The identities of the confessional poets are generic and, indeed, prototypically Romantic, in their embrace of emotional extremity and intensity. They also comply with the cultural expectations of confession (religious and psychoanalytic) in conveying intimacy and, by offering their private selves for public consumption, conscientiously embrace the frisson of transgression. Nevertheless, confessional poems, as Lerner puts it, have 'an aesthetic reason' for being written (66). They are composed from a rich, institutionally acquired knowledge of the domain, and often in direct response to the work of fellow poets in the confessional school.[7] The poetry is also motivated by professional ambition for success and recognition from the gatekeepers in their domain – for publication and acclaim, and even monetary reward. As Hughes poetically confesses of his and Plath's use of life in their work as young poets, 'we still weren't sure we wanted to own / Anything. Mainly we were hungry / To convert everything to profit' (1125).[8]

Such points are all germane to a theory of creativity that shifts the emphasis from individual psychology to what Glăveanu calls 'cultural psychology' (*Distributed Creativity* 8) – from a 'He-paradigm' or masculine 'paradigm of ... the exceptional creator' (7) to a 'We-paradigm' (8), which understands creativity as a '*distributed, dynamic, socio-cultural and developmental phenomenon*' (2, original emphasis). However, what I want to bring into focus here is how every creative act is also shaped by its materials, drawing attention to how the self that materialises through the autobiographical artefact is always other than itself. In what follows, I examine how confessional poetry emerges as a result of the writer making 'a more or less conscious decision to *share agency* with the object and follow its lead at different moments within the process' (Glăveanu *Distributed Creativity* 60, original emphasis). This mediation of the creative act by the technologies of poiesis – which includes the affordances and constraints specific to orality but also print, the parameters of poetic tradition, and the generative craft of rhythm and rhyme that Keats described as 'the wings of poesy' (60) – has implications for the agency of the autobiographical subject in the emergent event of life writing. Indeed, it is the powerful agency of the materials involved in the writing act that offers a different explanation for the lacuna in the poet's subjectivity metaphorically represented by 'madness'.

A confession about poiesis

In *Affective Disorder and the Writing Life*, a collection of autoethnographic essays by writers identifying as mentally ill, Stephanie Stone Horton upholds a literal connection between creativity and psychopathology and, evoking both Plath's poetry and her own autobiographical practice, argues that 'writing is like opening a vein' (3). I have no issue with the significance of embodiment and emotionality when it comes to poiesis, which I have attended to and theorised elsewhere ('After Romanticism' and 'Dissanayake's "Motherese"'). However, Stone Horton's representation of autobiographical poiesis as an organic event of individual expression is thoroughly Romantic. It recalls the mythopoeic work of the biopic *Sylvia*, in which, as David McCooey notes, the film 'conflates the body of work with the body' (293), especially at the film's conclusion when Plath (Gwyneth Paltrow) has committed suicide and left her manuscript of poems on the kitchen bench. Intercutting scenes show Hughes (Daniel Craig) 'kissing Plath's brow and touching the cover of *Ariel*' (293) as if they are interchangeable. However, even if we agree that a confessional poem is invested in embodied or emotional communication, it is certainly 'not an expression of emotion in exactly the same way as is a spontaneous change in facial expression' (Weisberg 256). A confessional poem, in other words, is not a frown or a smile or an act of blood-letting (to take literally Stephanie Horton's metaphor for creativity) – and this is because the domain-specific, culturally steeped and powerful technologies of poiesis play a significant role in the mediated emergence of the autobiographical poem and, importantly, of the autobiographical subject who is materialised therein.

There is a general tendency to ignore technology in favour of the romance of the self-expressive auteur when it comes to creativity, even if the aesthetic object in question is ostentatiously mediated, as in the case of photography or film, or even architecture and design, where the affordances and constraints of materials, and their cultural histories, are so obviously central to what evolves. The technologies of writing seem to be especially invisible. This is paradoxical when it comes to poetry, given how it calls attention to itself as a special form of language – through devices such as rhyme, metre, diction, its shape on a page – in a way that seems counter to the illusion of verisimilitude. However, according to Jay Bolter, the medium of writing, from the oral to the chirographic to the computer screen, has consistently been figured as a 'metaphor for the human mind' so that it has always been 'difficult for a culture to decide where thinking ends and the materiality of writing begins, where the mind ends and the writing space begins' (13). Language is, after all, what we think with; it is 'the principal embodiment of thought' (Bolter 189). It is also the case that, as N. Katherine Hayles suggests, we develop a 'technological unconscious' (134) through our habitual usage of materials that can render them inconspicuous. She explains how computing devices, for example,

> rapidly become integrated by way of proprioceptic, haptic, and kinesthetic feedback loops into the mind-body, so that agency seems to flow out of the hand and into the virtual arena in which intelligent machines and humans cooperate to accomplish tasks and achieve goals. (137)

However, as Bolter argues, writing is always a mediated event, even if it is achieved via older technologies: 'Despite its apparent immediacy … oral poetry is no more natural than writing, just as writing with pen and paper is no more natural, no less technological, than writing on a computer screen' (17).

In turning our attention to the mediation involved in the act of poiesis, let us begin by noting how orality and textuality introduce different constraints and affordances or, in other words, different forms of agency to the creative event. Of course, that agency is always impossible to separate from the contexts of the technology's cultural and historical usages, because objects 'gain intentionality in light of the action and intention of their makers, users, perceivers' (Glăveanu *Distributed Creativity* 60). However, 'the reverse is also valid: people become intentional actors precisely because they "confront" the intentions inscribed or discovered in objects' (60). Thus creativity is dependent upon a '*co-constructed and dynamic agency* fostered within person-object relations' (62, original emphasis).

Oral poiesis, for example, is a traditional rhythmic and formulaic technology of composition developed to aid memorisation and engagement prior to literacy. It relies on an existing network of stories, sustained through repetition and recall. Performance poets mobilising this technology today continue to produce poems enabled by familiar or popular content, rhyme, metre, repetition, a rhapsodic tenor, etcetera. The techne of an oral poetic tradition has by no means lost its power for page poets, including those of the confessional mode; the agency of oral techniques of composition continues to be apparent in their remediation into the chirographic and print. The craft of oral poetry, for example, is responsible for the formal arrangement of a poem into distinct lines and stanzas. Rhyme and metre also continue to generate words in ways that remain central to the creative act – something apparent in a poem such as Plath's 'Daddy', which is famously driven by a nonsense-like rhyme scheme: 'do'; 'shoe'; 'Achoo'; 'you'; 'blue'; 'Ach du'; 'Jew'; 'gobbledygoo' (54–5). However, textuality also introduces new affordances, allowing freedom from the demands of the mnemonic and the communal, enabling self-conscious revisionism and greater experimentation, and facilitating the personal content and free verse forms specific to the modern age. Indeed, writing and private habits of inscription provide poets with the capacity to perceive and respond to the inert marks on the page as evidence of their subjectivities in ways that undoubtedly facilitated the dominant development of modern personal lyric poetry – distinct from the 'resonant impersonality' of its ancient form (Lerner 46). The fact that these marks came to be set and bound in print, with the phenomenon of authorship 'sanctioned by copyright law' (Bolter 79) is also inseparable from the idea of the ontologically stable autobiographical self – from the ways in which, in relation to confessional poetry, the body of a poet can be identified with a poetry manuscript.

If we return our focus to the compositional process, however, the making visible of poiesis as technology that is enabled by textuality arguably only intensifies the effect of that technology.[9] Thus, the agency of the lyric poet is paradoxically compromised at the historical moment of its assertion. For the confessional poet, as for any other page poet, the making apparent of words and stanzaic shapes on the page engenders an acute awareness of aesthetic patterning, calling attention to line length, formal shape, and the emergence of repeating words and motifs. These properties have implications for content, with the length of lines and stanzas suggesting the scope or intensity of the event or emotion to be explored, for example, while emerging words and metaphors imply or constrain ideas or memories to be developed. The emerging and visible properties of the poem also trigger what Jonathan Lethem describes as an intertextual 'cryptomnesia' (59), recalling generic and specific arrangements and images from other half-forgotten poems, so that 'appropriation, mimicry, quotation, allusion, and sublimated collaboration consist of a kind of sine qua non of the creative act' (61) – and so that the agency of the self in the act of poiesis must be viewed as profoundly distributed.

Key to this phenomenon is Timothy Clark's conceptualisation of composition as 'mediated by self-reading' (15), which highlights how creative *reading* is an integral and underestimated component of creative *writing*.[10] As Clark explains, the writing 'is no sooner written than read' (19), so that the words on the page immediately begin suggesting 'unexpected directions for the text' (19). Like Paul de Man before him, Clark is aware of the implications for autobiography: 'Insofar as the space of composition embraces the ventriloqual effects of received literary codes and their constraints ... *then the agent or subject of composition cannot be simply identified with the empirical subject*' (24, original emphasis). Indeed, according to Clark, it is the agency of the writing technology that explains the 'hiatus in the structure of subjectivity' (19) that lyric poets and scholars of poiesis, in particular, have so often explained 'as the manifestation of hidden "depths" of the mind' (29). For Clark, 'effects that seem psychic, internal or psychological are often determined by the material parameters of composition' (40).

Certainly, I have experienced the agency of the materialised technologies of poiesis in precisely these ways when writing confessional poetry. A developing internal rhyme scheme often surprised me with words that changed the direction or content of the poem. An image that emerged on the page, and that I subsequently experienced as aesthetically charged, spontaneously informed the elaboration of a metaphorical consciousness that itself elicited autobiographical details. Glăveanu describes the dynamic process of creativity thus: 'the object at each moment gives material form to – and, in doing so, responds to and changes – the intentions, goals and representations of the creator' (*Distributed Creativity* 60). However, even before my confessional poems materialised, even before I began composition, I was profoundly conscious of the agency of the confessional genre – with all its psychoanalytic and religious as well as poetic history – which I had chosen to deploy. The confessional genre essentially provided a constraint that, as Patricia Stokes writes, 'turns an initially ill-defined problem into a well-defined one' (130) and 'provides you both with the things you can *work with* and the things you can *work against*' (131, original emphasis). The techne of genre narrowed the scope of my autobiographical experience to be 'confessed' to that marked by psychological intensity; it permitted, even demanded, a heightened or melodramatic tone, which also had an effect on the choice of autobiographical detail and the emergence of image schemes. Thus, my sequence of confessional poems are not representative of my identity more generally – and, indeed, the same might be said of the confessional poems of my precursors. My poems focus on a brief but particularly challenging time, when my husband's life-threatening illness brought about a breakdown of sorts for me, triggering the revisitation of memories and a mindset from a traumatic past. Neither are my poems generated by 'madness' – my period of mental unwellness having well and truly passed by the time I began composing – but, profoundly, by craft.[11] Explicitly acknowledging the importance of the genre of confessional poetry, most of my poems take their titles from other confessional poems, are addressed to the relevant poets, and refer to biographical details or lines from their work. This introduced another constraint, which functioned as another technology of affordance. For example, a poem called 'Argument' is addressed to Elizabeth Bishop (who was closely associated with the confessional poets if not identical with that school) and takes its title from one of her poems. Bishop's Brazilian connection then provoked me to recall an episode in South America from my own life in the subject of the poem.

My sequence of confessional poetry begins with a poem called 'Cruel', which is addressed to my husband, who is also a poet. When I began writing this poem – on a computer that always allows my work a heightened sense of provisionality owing to the enhanced ease of revision – the disorderly appearance of the unstructured words and cryptomnesiac memories of the formal appearance of sonnets led to the poem's rendering into two stanzas of equal length. This evolution of formal discipline in turn imposed a discrete and episodic approach to my narration of life events. While not strictly following the line length of the conventional sonnet, the second stanza of each poem also developed a kind of volta, which turns – or returns – the poem to an earlier image to provide an aesthetic sense of narrative coherence. These features of the poem emerged in unconscious and dynamic ways, as intuitive responses to the agency of the inchoate forms on the computer screen and the potency of the domain of poetry in which I have long been immersed.

The craft of poetry thus elicited what is an ultimately profoundly aestheticised confession. This is not to deny my own agency as poet and autobiographical subject, which was indeed often in conflict with the affordances and constraints of the materials – in '*tension* in relation to what the material support "allows" or "forbids"' (Glăveanu *Distributed Creativity* 57). However, the act of poiesis that led to the emergence of my confessional poems can at best be understood, as the poet Dean Young describes it, as 'the manifestation of choices in a charged field' (33).

As Elizabeth Gregory writes in 'Confessing the Body', 'poets ... have long been part of the active penumbra of poetry, throwing shadows around the work that inform its reception' (36). Given that confession has provided a culturally sanctioned ritual for producing truth since the Middle Ages and that the lives of the confessional poets have been so sensational, it is no wonder that confessional poems, in particular, have become identical with their creators. Confessional poetry does, of course, shed light on the lives of confessional poets; my poems, like those of the confessional poets that come before me, rely on the kinds of autobiographical details that can be biographically substantiated in ways that satisfy criteria for life writing. At the same time, however, those poems reveal autobiography's limits. They come into being only through a profound and necessary unsettling of the autonomy of the autobiographical subject in an engagement with the writing craft. In fact, the agency of the autobiographical writer exists only in relation to the writing materials that enable her art.

Notes

1. "Cruel." *Kenyon Review* (US) XXXIX. 2 (2017): 86–87; "Shattered Head." *Island* 148 (2017): 45; "Argument." *Best Australian Poems 2016*. Ed. Sarah Holland-Batt. Collingwood: Black Inc, 2016. 160–1; "Daddy." *Island* 146 (2016): 96; "Nox." *Cordite*, 1 August 2016; "Déjà vu." *Australian Book Review*. January–February 2016: 14; "Knife." *The Best Australian Poems 2015*. Ed. Geoff Page. Collingwood: Black Inc, 2015. 32–3.
2. My intention here needs to be distinguished from related approaches to confessional poetry, such as Jacqueline Rose's *The Haunting of Sylvia Plath*, which focuses on reading and on how the hermeneutic ambiguity of the poems, as 'literature', complicate any closed biographical readings. My focus is on writing and on how the materials or technologies of the poem

have an agency that complicates biographical readings. In other words, my concern is not with the problem of mimesis but of poiesis. Susan Van Dyne's *Revising Life: Sylvia Plath's* Ariel *Poems* comes closer to what I am attempting, attending to the material practices of Plath as poet, which include her preference for the 'pink bond of Smith College memorandum paper' (with its stamp of cultural authority) and her drafting of the *Ariel* poems 'on the reverse of several of Hughes' manuscripts ... and on the reverse of an edited typescript of *The Bell Jar*' (8). However, Van Dyne's intention is to ultimately redeem a feminist 'agency' (3) for Plath as someone who consciously revised her life and personae through her poems. Mine is to interrogate the agency of all writers, who inevitably share creative agency with the materials or technologies of their craft.

3. For a discussion of the dynamic relationship between Romanticism and psychoanalysis, see Joel Faflak's *Romantic Psychoanalysis*, in which he argues that Romanticism '*invents* psychoanalysis' (15, original emphasis) through its interest in 'the spectre of the Enlightenment's absent psychosomatic body' (26) and its use of lyric poetry as a 'type of close self-reading' (24) approximating the 'analytic scene' (25).
4. When Berryman killed himself, W.H. Auden reportedly joked that his suicide note read '"Your move, Cal"' (quoted in Redfield Jamison 213). (Cal was the affectionate name for Lowell used by his friends.) In a similarly cynical vein, John Moffitt argues that creative practitioners materially benefit from conforming to the figure of the 'artist-priest-prophet' (258) given that it comes with 'a complementary syndrome, "Art-Worship"' (267). As evidence of such a thesis, one might draw attention to the different statuses of the contemporaneous poets Anne Sexton (who committed suicide) and Maxine Kumin (whose life was much more stable), as Jean Tobin does, arguing that, 'if a disturbed life is part of the public's definition of *poet*, the public will be predisposed to pay more attention to poets with disturbed lives' (46, original emphasis).
5. As Diane Wood Middlebrook argues, Lowell's 'pedigree' – not to mention that of his other middle-class contemporaries – was also important in giving 'rank and station to confessional poetry' (640).
6. Suffice it to say here that the notion of 'madness', being historically constructed (as Michel Foucault and the constantly changing Diagnostic and Statistical Manual of Mental Orders show), and the notion of creativity, which is likewise contingent, are each so vague that linking them 'is like trying to get two clouds to stick together' (Schlesinger "Building Connections" 64). Martin Lindauer, Judith Schlesinger ("Building Connections", *The Insanity Hoax*), and Arne Dietrich usefully interrogate the methodologies of the most commonly cited studies that affirm a link between mental illness and creativity, showing how they are compromised by small samples, retrospective diagnosis, literary analysis, and confirmation bias. These studies can be contrasted with the Harvard psychiatrist Albert Rothenberg's 25-year study of creative subjects, involving more than 2000 hours of interviews, which revealed only the common trait of 'motivation' (8).
7. Lowell taught and mentored W.D. Snodgrass, Sexton and Plath, and was good friends with Berryman. Plath and Hughes responded to each other's poems, as is well known, and Sexton wrote in response to work by Snodgrass (Lerner 53). It is notable that, in the biopic *Sylvia*, as David McCooey notes, this wider network of poets is absent, 'thus emphasising the "natural" or "pathological" relationship between Plath and vision' (295), a relationship so often mobilised in relation to the confessional poets. However, according to a cultural theory of psychology such as the one being defended here, a confessional poem cannot be understood simply as the manifestation of an individual mind; it emerges from and has value only within a network of creative practice.
8. It is an impulse for which the poem, figuring the selling of a life in relation to the selling of daffodils, expresses a poignant regret – while managing to blame Plath for the violation: 'your scissors remember. Wherever they are' (1126).
9. This is evident in the modernist and postmodern work of the L-A-N-G-U-A-G-E poets and others such as Kenneth Goldsmith and Caroline Bergvall who emphasise textuality, rejecting

the expressive lyric in favour of a performative dissolution of 'their putative authority into … the materials of poetry' (Brown 24).

10. Similarly, the poet Mark Yakich encourages his student writers to think of creativity as revision – a process antithetical to the *furor poeticus* – and of revision not as 'correction' but as 'opening up the possibilities of what's already on the page' (154). He writes: 'Amateur writers expect their lines to descend from on high or to be whispered in their ear by a muse. Nothing could be further from what happens in practice. For writing is all practice' (196). Writing, in other words, materialises. The self, too, materialises in autobiographical writing.
11. As Plath put it, challenging the organic connection between 'madness' and poiesis, '[w]hen you are insane, you are busy being insane – all the time … When I was crazy, that's all I was' (quoted in Flaherty 66).

Disclosure statement

No potential conflict of interest was reported by the author.

References

Bolter, Jay David. *Writing Space: Computers, Hypertext, and the Remediation of Print*. London: Routledge, 2011.

Brain, Tracy. "Dangerous Confessions: The Problem of Reading Sylvia Plath Biographically." *Modern Confessional Writing: New Critical Essays*. Ed. Jo Gill. Abingdon, Oxon: Routledge, 2006. 11–32.

Brockmeier, Jens and Donal Carbaugh. "Introduction." *Narrative and Identity: Studies in Autobiography, Self and Culture*. Eds. Jens Brockmeier and Donal Carbaugh. Amsterdam: John Benjamins, 2001. 1–22.

Brockmeier, Jens and Rom Harre. "Narrative: Problems and Promises of an Alternative Paradigm." *Narrative and Identity: Studies in Autobiography, Self and Culture*. Eds. Jens Brockmeier and Donal Carbaugh. Amsterdam: John Benjamins, 2001. 39–58.

Brown, Nathan. *The Limits of Fabrication: Materials Science, Materialist Poetics*. New York: Fordham University Press, 2017.

Bruner, Jerome. *On Knowing: Essays for the Left Hand*. Cambridge, MA: The Belknap Press, 1980.

Clark, Timothy. *The Theory of Inspiration: Composition as a Crisis of Subjectivity in Romantic and Post-Romantic Writing*. Manchester: Manchester University Press, 1997.

Dietrich, Arne. *How Creativity Happens in the Brain*. Houndmills, Basingstoke: Palgrave Macmillan, 2015.

Faflak, Joel. *Romantic Psychoanalysis: The Burden of the Mystery*. Albany, NY: State University of New York Press, 2008.

Flaherty, Alice. *The Midnight Disease: The Drive to Write, Writer's Block, and the Creative Brain*. New York: Houghton Mifflin, 2005.

Foucault, Michel. *Madness and Civilization: A History of Insanity in the Age of Reason*. Trans. R. Howard. London: Routledge, 1989.
Gill, Jo. "'Your Story, My Story': Confessional Writing and the Case of *Birthday Letters*." *Modern Confessional Writing: New Critical Essays*. Ed. Jo Gill. Abingdon, Oxon: Routledge, 2006. 67–83.
Gill, Jo. "Introduction." *Modern Confessional Writing: New Critical Essays*. Ed. Jo Gill. Abingdon, Oxon: Routledge, 2011. 1–10.
Glăveanu, Vlad-Petre. "Principles for a Cultural Psychology of Creativity." *Culture and Psychology* 16.2 (2010): 147–163.
Glăveanu, Vlad-Petre. *Distributed Creativity: Thinking Outside the Box of the Creative Individual*. New York: Springer, 2014.
Glăveanu, Vlad-Petre. "Craft." *Creativity – A New Vocabulary*. Eds. Vlad-Petre Glăveanu, Lene Tanggaard and Charlotte Wegener. Houndmills: Palgrave Macmillan, 2016. 28–35.
Gregory, Elizabeth. "Confessing the Body: Plath, Sexton, Berryman, Lowell, Ginsberg and the Gendered Poetics of the 'Real'." *Modern Confessional Writing: New Critical Essays*. Ed. Jo Gill. Abingdon, Oxon: Routledge, 2011. 33–49.
Hayles, N. Katherine. *Electronic Literature: New Horizons for the Literary*. Indiana: University of Notre Dame Press, 2008.
Hughes, Ted. "Daffodils." *Collected Poems*. London: Faber and Faber, 2003. 1125–6.
Keats, John. "Ode to a Nightingale." *The Complete Poems*. Harmondsworth: Penguin, 1981 (1819). 346–8.
Kyaga, Simon. *Creativity and Mental Illness: The Mad Genius in Question*. Houndmills, Hampshire: Palgrave Macmillan, 2015.
LaTour, Bruno. *Reassembling the Social: An Introduction to Actor-Network-Theory*. Oxford: Oxford University Press, 2005.
Lerner, Laurence. "What is Confessional Poetry?" *Critical Quarterly* 29.2 (1987): 46–66.
Lethem, Jonathan. "The Ecstasy of Influence." *Harper's Magazine*, February 2007: 59–71.
Lindauer, Martin S. "Are Creative Writers Mad? An Empirical Perspective." *Dionysus in Literature: Essays on Literary Madness*. Ed. Branimir M. Rieger. Bowling Green, OH: Bowling Green State University Press, 1994. 33–48.
Lucas, Rose. "Gifts of Love, Gifts of Poison: Anne Sexton and the Poetry of Intimate Exchange." *Poetry and Autobiography*. Eds. Melanie Waters and Jo Gill. London: Routledge, 2011. 43–57.
McCooey, David. "Visions and Sensations: Poets on Film." *Literature and Sensation*. Eds. Anthony Uhlmann, Stephen McLaren, Paul Sheehan and Helen Groth. Newcastle upon Tyne: Cambridge Scholars Publishing, 2009. 290–99.
Moffitt, John. *Inspiration: Bacchus and the Cultural History of a Creation Myth*. Leiden: Brill, 2005.
Plath, Sylvia. "Daddy." *Ariel*. London: Faber and Faber, 1990 (1965). 54–6.
Redfield Jamison, Kay. *Robert Lowell. Setting the River on Fire: A Study of Genius, Mania and Character*. New York: Alfred A Knopf, 2017.
Rose, Jacqueline. *The Haunting of Sylvia Plath*. London: Virago, 1991.
Rothenberg, Albert. *Creativity and Madness: New Findings and Old Stereotypes*. Baltimore: The Johns Hopkins University Press, 1990.
Schleifer, Ronald. *Intangible Materialism: The Body, Scientific Knowledge, and the Power of Language*. Minneapolis: University of Minnesota Press, 2009.
Schlesinger, Judith. *The Insanity Hoax: Exposing the Myth of the Mad Genius*. Ardsley-on-Hudson, NY: Shrinktunes Media, 2012.
Schlesinger, Judith. "Building Connections on Sand: The Cautionary Chapter." *Creativity and Mental Illness*. Ed. James C. Kaufman. Cambridge: Cambridge University Press, 2014. 60–75.
Smith, Sidonie and Julia Watson. *Reading Autobiography: A Guide for Interpreting Life Narratives*. Minneapolis: University of Minnesota Press, 2010.
Sontag, Susan. *Illness as Metaphor: Aids and its Metaphors*. London: Penguin, 1991.
Stokes, Patricia. *Creativity from Constraints: The Psychology of Breakthrough*. New York: Springer, 2005.

Stone Horton, Stephanie. "'What Ceremony of Words Can Patch the Havoc?' Composition and Madness." *Affective Disorder and the Writing Life: The Melancholic Muse*. Ed. Stephanie Stone Horton. Houndsmills, Hampshire: Palgrave Macmillan, 2014. 2–23.

Takolander, Maria. "After Romanticism, Psychoanalysis and Postmodernism: New Paradigms for Theorising Creativity." *TEXT* 18.2 (2014). <http://www.textjournal.com.au/oct14/takolander.htm>.

Takolander, Maria. "Dissanayake's 'Motherese' and Poetic Praxis: Theorising Inarticulacy and Emotion." *Axon: Creative Explorations* 4.1 (2014). <http://www.axonjournal.com.au/issue-6/dissanayake's-'motherese'-and-poetic-praxis>.

Takolander, Maria. "From the 'Mad' Poet to the 'Embodied' Poet: Reconceptualising Poetic Creativity through Cognitive Science Paradigms." *TEXT* 19.2 (2015). <http://www.textjournal.com.au/oct15/takolander.htm>.

Takolander, Maria. "Argument." *Best Australian Poems 2016*. Ed. Sarah Holland-Batt. Collingwood: Black Inc., 2016. 160–1.

Takolander, Maria. "A Dark/Inscrutable Workmanship: Shining a Scientific Light on Poiesis," *Axon: Creative Explorations* C1, 2016. <http://www.axonjournal.com.au/issue-c1/darkinscrutable-workmanship>.

Takolander, Maria. "Cruel." *Kenyon Review* (US) XXXIX. 2 (2017): 86–87.

Tobin, Jean. *Creativity and the Poetic Mind*. New York: Peter Lang, 2004.

Van Dyne, Susan. *Revising Life: Sylvia Plath's Ariel Poems*. Chapel Hill: University of North Carolina Press, 1993.

Waters, Melanie and Jo Gill. "Introduction." *Poetry and Autobiography*. Eds. Melanie Waters and Jo Gill. London: Routledge, 2011. 1–24.

Weisberg, Robert. "Problem Solving and Creativity." *The Nature of Creativity: Contemporary Psychological Perspectives*. Ed. Robert J. Sternberg. Cambridge: Cambridge University Press, 1988. 148–76.

Wood Middlebrook, Diane. "What was Confessional Poetry?" *The Columbia History of American Poetry*. Ed. Jay Parini. New York: Columbia University Press, 1993. 632–49.

Yakich, Mark. *Poetry: A Survivor's Guide*. New York: Bloomsbury, 2016.

Young, Dean. *The Art of Recklessness: Poetry as Assertive Force and Contradiction*. Minneapolis: Graywolf Press, 2010.

REFLECTION

I Guess What You Say is True

Oliver Driscoll

ABSTRACT
This work of non-fiction struggles with the technical and ethical difficulties of representing the former Yugoslavia, an area that has experienced deep trauma and that is at a remove from the author's experience. The author-narrator confronts the challenges of representing others' trauma in an oblique way: focusing on his own life in Melbourne and his responses to Yugoslavia's past; critiquing narrative forms and particular literary works that represent the trauma, namely those of the Bosnian-American writers Aleksandar Hemon and Semezdin Mehmedinović; and acknowledging the vexed framework of the creative-writing PhD for which the reflection was produced.

Part 1

For some time I'd been worried about a dull pain in the centre of my back, around one of the vertebrae, which seemed to feel unusually large. I'd tried to count up and count down to which vertebrae it was. Unsurprisingly, however, I'd been unable to do so. My doctor asked how long I'd had this pain. I said, eight months, since around the time I went overseas. This was inaccurate, firstly because I'd travelled overseas closer to 11 months earlier, and secondly because I didn't know when it appeared.

My doctor asked, on a scale of one to ten, how severe the pain was. I said, one or two, if that. My concern, I said, was for the longer term. I tried to explain to him that I was worried that it was almost painless because of its seriousness.

In October the previous year, I'd flown with my partner, Jamie, to Sarajevo. There, we'd lived in an apartment with a dark-timber sloping ceiling rented through Airbnb for four months. Sarajevo is sunken down between green hills. I'd never been to Switzerland, but imagined it to feel like Sarajevo, though without the prayer calls, the tall thin minarets, the occasional spray of bullet holes across buildings, and the funny little patches of in-fill terracotta on the sides of housing blocks where they'd been slammed with mortar.

In the apartment with the dark timber sloped ceiling, I had trouble sleeping, in part because I frequently do, in part because of the dark brown moths that would hide against the ceiling during the day but that at night would scuttle across the glass doors that opened onto our balcony trying to get to the light of the Twist Tower – a tall spiralling commercial building down on the flat near the station – and in part because of the light from the tower itself.

The Twist Tower was the only thing of its kind in that part of town. Despite our apartment being up in the hills and the tower down on the flat, it always felt close; it was hard to take photos from our balcony without including the tower. At night it was impossible to get away from the light.

I would sometimes have trouble falling asleep and I would sometimes wake early, unable to fall back asleep. I tried sleeping in the kitchen, which was separate to the living and sleeping space, but there the opaque skylight would glow with the light from the Twist Tower. Often, in the early hours, unable to fall back asleep and being too tired or disappointed to read, I would lie there trying to think dull but not tedious thoughts.

Very occasionally, I would get out of bed, put on my navy-blue puffer jacket, pack an umbrella, my camera and a few plastic bags into my satchel, and head out to take photographs of the city. I am not a good photographer, and try to get by on volume. I would walk around the hills, wary of stray dogs, though needlessly as they were not even curious. At the time, I was not experiencing great discomfort in my back, not even a one or a two in terms of pain, but I would occasionally on these walks, if the morning was cold enough, and if I was out long enough, lose strength in my arms. When I could no longer twist my camera's lens in order to turn it on or off, there was nothing I could do but walk home. In any case, I would try to make sure I was home by 9 am to have breakfast with Jamie.

When the problem with my arms was particularly bad, I would ring the doorbell, so as to not have to struggle with the key. I would then have to downplay the loss of strength in my arms to Jamie, and think of an excuse for having used the doorbell.

I was in Sarajevo in relation – that is perhaps the best way of putting it, *in relation* – to my creative writing PhD. Whenever I told people I was a creative writing PhD candidate, I could see them trying to work out if this meant I was unusually intelligent, or unusually lacking in intelligence, and whether it would be impolite to ask further questions, such as, what is a creative writing PhD, why do they exist, and what do I intend to do with the rest of my life?

Sometimes I would tell people I was doing a literature PhD in part because it sounded academic in a way that creative writing PhDs don't, and in part because it meant only having to explain the critical part of my dissertation, which was on Aleksandar Hemon, a writer from Sarajevo who now lives in Chicago.

I had since discovered Bosnian writers who were more interesting and more challenging than Hemon, though unlike Hemon they didn't write in English. As I don't speak what is sometimes called Serbian, sometimes Croatian, sometimes Serbio-Croatian and sometimes Bosnian, I could only read them in translation and so had been advised by my supervisor not to focus too heavily on them.

I'd had many titles for my PhD, all of which were partially self-deprecating and partially attempts to sound serious about my research. The title at the time of the doctor's visit was *Trauma and the Imaginary: Aleksandar Hemon, and the Balkans as a Place Where Death Occurs*. The most recent change was substituting the word 'happens' for 'occurs'.

When I was seeing the doctor about my back in Melbourne, Jamie and I were sitting for friends in a house in Fitzroy that was nicer than one we'd ever be able to afford to live in again. We were taking care of their two Burmese cats, Trotter and Otter, who were collectively the size and mass of three, if not four, Sarajevo cats. I went to a doctor in St Kilda because that's where I'd lived when I first moved to Melbourne some years earlier.

My doctor was short and had silver hair. He was gentle and quietly charming – the kind of doctor you would invite to your book launch. He would ask me things like the etymology of medical terms, as though we were two clever, learned men. Unfortunately, if a doctor said a word like 'etymology' my brain tended to stop functioning. 'Etymology' would lose its distinctiveness, quietly vibrating against words like epistemology and ontology. I wouldn't hear the word for which an etymology was being sought.

Before leaving to take the tram from Carlton Gardens to my doctor's practice in St Kilda, I packed an umbrella and *Two Lives* – Janet Malcolm's dual biography of Gertrude Stein and Alice B. Toklas – into my satchel. When I mentioned to Jamie that I'd been reading Malcolm, she said that she found her frustrating, that her 'I' was too general and impersonal, and that her writing was broken up with enormous quotations. I picked up *Two Lives* after reading her *Paris Review* 'Art of Nonfiction' interview with Katie Roiphe. Roiphe asked Malcolm about this tendency to heavily use quotations.

> Interviewer
>
> I notice in your answers to my questions a kind of collage element. You will often paste in long quotations, and that is also true of your nonfiction and criticism. Can you explain your attraction to this technique?
>
> Malcolm
>
> Well, the most obvious attraction of quotation is that it gives you a little vacation from writing – the other person is doing the work. All you have to do is type. But there is a reason beyond sloth for my liking of quotations at length. It permits you to show the thing itself rather than the pale, and never quite right, simulacrum that paraphrases it. For this reason I prefer books of letters to biographies. I am tempted to quote myself on this subject … but you have made me feel self-conscious, maybe even a little guilty, about this practice, so I will resist the impulse.

I got off the tram across from Luna Park and walked to the medical centre. Inside, there was a woman sitting close to the counter, leaning toward the receptionist. She asked how much longer she would be required to wait. The receptionist said, 'Well, you were here early, and she is running late, so perhaps another 20 minutes.' The woman asked if she could leave and come back. The receptionist said, yes, so long as she could be reached by mobile. The woman didn't move – she stayed there, leaning toward the receptionist.

I came upon Hemon's work while writing my masters thesis, for which I was planning to write about the concept of testimony in relation to E.L. Doctorow's *Lives of the Poets: Six Stories and a Novella*. I'm poor at organising my research material, and while *Lives of the Poets* remained in the thesis, the word 'testimony' never appeared. I borrowed a great many books from the library, more or less related to testimony, including *Testimony after Catastrophe: Narrating the Traumas of Political Violence* by Stevan Weine.

Building on the theories of Mikhail Bakhtin, Weine writes that following manmade or natural catastrophes, measures should be in place to prevent a dominant narrative from adulterating or muddying the unique accounts of individuals. Weine discusses Hemon's story, 'A Coin', which I found through ProQuest, having originally been published in the *Chicago Review* in the winter of 1997. The story is epistolary: we see the letters between a man who – like Hemon – has escaped to America, and a woman in siege-time Sarajevo. There is a long delay between when her letters are sent and when he receives them; he cannot know if she's still alive by the time he reads them.

There was a poster in the medical centre that the woman would have been blocking had she not been leaning toward the receptionist. I remembered now that I'd previously sat here looking at the poster with some unease. Then, as now, I hadn't been sure whether the wording made sense or not. It read: 'Pap test. A little awkward for a lot of peace of mind.' I thought it should read, 'A little *bit* of awkward*ness*', but the more I read the poster, the less certain I was. I would try other formulations and soon nothing would seem entirely clear.

I sat down in the waiting room and took out *Two Lives*. I couldn't concentrate. The medical centre was the only place I encountered daytime television. This time there was a reality show on with three women and one man. The women took turns hosting a dinner for the man and the other two women in their houses or apartments, or in places we were told were their houses or apartments.

Before each dinner, the host would provide the others with a menu on shiny metallic paper. One of the women, now sitting in a park, waved the current host's menu around, as though trying to put it out. She said, *I didn't know custard on its own was a dessert.* In a dark apartment, after the custard, which seemed to be a success, despite the man not really eating his, the four of them played strip Twister. The women hadn't seemed to be wearing swimsuits under their clothes when they'd arrived, nor had they been carrying bags, but one of the women was now playing Twister in a bikini. The man was down to board shorts, which he also hadn't been wearing earlier. The host, sitting in a room, said to the camera that she was upset that she didn't have the opportunity to remove her clothes. The third woman refused to play.

When I first read Hemon's story 'A Coin', I thought it was superb, that it conveyed something of the war in Bosnia accurately and truthfully. When I read the story now, however, I no longer have that impression. Now when I read the story, I am very conscious of how it wants me to feel. It reads like fiction.

The deaths and brutalities in the story were, I suppose, the first I consciously read about from the war in Bosnia. As a teenager, I'd seen civilians – who we would perhaps now call *non-combatants* (or *enemy non-combatants*) – on televisions being shot at by snipers, or lying dead in the street wearing city clothes, or making their way through mountains from who knows where to who knows where while mortar exploded around them, shaking the camera and altering the voice of the reporter.

I have since watched too many hours of news footage from the war on YouTube for my memory of this earlier viewing to remain unchanged, but these images seemed to be on television for years. More likely, after Bosnia they disappeared, more or less, for a few years before similar images resurfaced, this time from Albania. At the time, I didn't think much of them. They were what the news looked like.

At the time, the words 'Sarajevo', 'Belgrade' and 'Mostar' had a very particular feel. I wouldn't know for a long time that the Balkans had an ability to open and shut, to at once be part of Central Europe, and at once an unknowable zone between Greece and Austria.

When I first read Hemon's 'A Coin', I was running a literary event in Melbourne called the Slow Canoe Readings. I asked an actor who I'd been in a postgraduate class with to read 'A Coin' at our next event. I sent her the PDF from ProQuest and she agreed. Unfortunately, however, the story was far too long. Our audience would become impatient, the actor said. She suggested we cut the paragraphs referring to the cats and dogs.

The actor said that although they were some of the best paragraphs in the piece, someone unfamiliar with the story wouldn't notice their absence. The story would not be disrupted and no ambiguity would be introduced. I thought there were perhaps many stories in which some of the best paragraphs could be struck out without the reader noticing.

The first of the paragraphs referring to cats and dogs, the first we removed, reads as follows:

> Sarajevo is a catless city. It is so because people couldn't feed them, or couldn't take them along when they were fleeing, or their owners were killed. Hence the dogs that couldn't be fed or taken along hunt them down and devour them. One can often see, among the rubble on the streets, underneath burnt cars, or stuck in sewers, cat carcasses, or cat heads with death grins, eyeteeth like miniature daggers. Sometimes one can see two or more dogs fighting over a cat, tearing apart a screaming loaf of fur and flesh.

Sarajevo today is not a catless city – it's a city of many stray cats, just as it is a city of many stray dogs. I don't know if cats and dogs have since made peace. Perhaps Hemon may have been incorrect to have blamed dogs for the absence of cats. It may have been humans who were hunting them. This would explain why they have since returned despite the continued prevalence of stray dogs. Or he may have been wrong about Sarajevo being a catless city altogether.

Semezdin Mehmedinović, one of the Bosnian authors who I found more interesting than Hemon, writes in his essayistic story 'Soviet Computer', published in part in the 'Casualty' issue of the *Massachusetts Review*:

> I was walking home in the dead of night, after curfew, taking the side streets so as to avoid the soldiers. I heard a sound behind me, and turning around I saw a hundred wild, faintly growling dogs following me. Fluorescent eyes in the dark, and growling that makes the blood run cold. Horrified, I continued walking – I had become the nape of my unprotected neck – in the direction of the nearest familiar house. I went into Hrvoje's yard and knocked on the door. The pack followed me into the yard and the growling turned into barking. Thankfully, the dogs don't know that I no longer have the strength to put up a fight. Hrvoje was glad to see me, but not nearly as glad as I was to see him. Sooner or later, somebody has to win. Either we will survive, or the wild dogs will devour us.

In the early months of my PhD, my supervisor requested that I meet with her every week to talk about my progress. She said it was a good idea for her to read everything I wrote. Her office was next to mine and I would frequently run into her in the corridor. If I wasn't seen in the corridor for a few days, she would ask me where I had been and whether I'd been having any trouble that I would like to discuss with her.

She spoke to me, at that time, as though we were two unusually intelligent people amongst less intelligent people. She told me to think about *form*. She said *form* so many times that I started to lose track of what it meant.

She said my writing was strong and good, that it was refreshing to see someone looking at authors and subject areas from beyond this country. She said, however, that if I'm interested in pain, violence and suffering, I should look nearer to home, at all the suffering that has gone on in Australia.

She gave me a short list of books to read, mostly by Aboriginal authors, and a longer list of books that I ought not to read, mostly by non-Aboriginal authors writing about Aboriginal people, though there were a few Aboriginal authors on this list also. She said, the only authors doing anything interesting in this country are Indigenous, and I should keep a diary and record my changing perspective on the matter. She said, you know, as a white guy. I said yes, I understood, it was a good idea, and I meant it.

My supervisor said I could do something really good, and that this is exactly *what we need.* She placed the palm of her hand down on the table on each of the last three words: *what – we – need.*

I agreed with my supervisor, or at least thought I did. I bought and borrowed the books on her list, but I couldn't bring myself to read them, at least not thoroughly and with real concentration.

My supervisor said that formal experimentation is crucial, that it's the only absolute, and that realism was a problem. Too many Australian writers were realists, she said, with a disdain I could see in her jaw and in her neck. She had a neck you wanted to look at. I asked her if I could quote her. Of course, she said.

I agreed with her that certain authors were good and that others were not, but I thought, probably wrongly, that her terminology was incorrect or misleading, that to her realism and bad writing meant the same thing. I thought her assessment had more to do with the idea of the works we discussed than the works themselves, as though particular forms or styles of writing could stand in for, or contain, particular political positions. I promised I would think more about form.

One morning in bed in Sarajevo, after the alarm had gone off several times, Jamie said, I guess when we're thinking about a writer finding their voice, we're talking about them finding that thing that sounds most like their natural voice, without ever having to expose too much. I said, yeah, I suppose so. I got out of bed, and she asked if I could pull the blinds open. I'd had a dream about an architectural drafting job that a friend had sent me. In the dream, the job title was 'director of light'.

Since encountering the deaths and brutalities in Hemon's 'A Coin', I'd read of countless more. For the creative component of my PhD, I was required to write 60,000 words. If I were to have collated the descriptions of all the deaths I had read about, restricting myself to the conflict in Bosnia alone from the years 1992 to 1996, I would have easily satisfied the word count. If I were to have extended this collage ahead into Albania, and then back into Croatia, and through Tito's time to the Second World War, and further back still, say, to the plains of Kosovo, I would have needed a far greater word limit, and far more sympathetic assessors than I could ever hope for.

At the general practice in St Kilda, my doctor eventually called me in and began gently percussing my spine. He said, you were going away somewhere interesting the last time I saw you. I said, Sarajevo. He said, ahh, beautiful. I had hardly slept the night before. I said, yes, I suppose so.

He said, well, not beautiful, I suppose. Isn't it in a valley? Are there any signs of the recent troubles? I asked whether he meant recent troubles as in the war, or as in the floods. He said, the war, and I said, there were, but actually it was a very beautiful city.

I was very tired, but tried to explain that one thing that was particularly beautiful about Sarajevo was that it looked and felt very European, but was predominantly Muslim. I'd had this discussion several times already, and at best I would have difficulty not sounding as though I could appreciate Islam so long as it was less Islamic and more European. I certainly couldn't do this then with the doctor. He told me, as he had done before I travelled, that his wife – the doctor in the next room for whom the woman had been waiting – was Macedonian.

My doctor went on to say that his wife – who he now referred to as Dr Andrews – had many clients from the Balkans and that the floods had brought them together. He smiled. I tried to say that it wasn't altogether surprising, that the widely held belief about the conflict being based on ancient ethnic hatred was misleading. I said it was more accurate to say the war was run by thugs as a pretext to grab state-owned property, factories, everything. I should have stopped talking, though. The war was much messier than this.

Before the doctor's appointment, I'd read that the floods had exposed mass graves. Talking to him, I felt as though I was collaging together undigested material that I didn't altogether believe in. Though I didn't not believe in it either.

An alternate version of 'A Coin' appears in Hemon's story collection *The Question of Bruno*, though the differences between this version and the one that appeared earlier in the *Chicago Review* are not significant. The story ends with the woman in Sarajevo writing about running across intersections that snipers are known to frequently fire upon. This is how the final paragraph reads in the *Chicago Review* version:

> I've run from Point A to Point B hundreds of times, and the feeling is always the same. But I've never had it before. I suppose it is this high pressure of excitement that makes people bleed away so quickly. I saw deluges of blood coming out of thin bodies. I saw streams of blood spouting from children as if from fountains. But once you get to Point B everything is quickly gone, as if it never happened. You pick yourself up and walk back into your besieged life, happy to exist. You brush a wet curl from your forehead, inhale deeply, and put your hand in your pocket, where you may or may not find a worthless coin – a coin.'

In the later version, the one that appears in *The Question of Bruno*, the paragraph reads:

> I've run from Point A to Point B hundreds of times and the feeling is always the same but I've never had it before. I suppose it is this high pressure of excitement that makes people bleed away so quickly. I saw deluges of blood coming from svelte bodies. A woman holding on to her purse while her whole body is shaking with death rattle. I saw bloodstreams spouting out of surprised children, and they look at you as if they'd done something wrong – broken a vial of expensive perfume.

I bristle at Hemon's 'svelte bodies', though I understand that the supposed inappropriateness is intended. It suggests, at once, the letter writer's unfamiliarity with English, and that they're not in their right mind – they're noticing sexual physicality when they ought not to be, given the moment of great stress and threat. Nevertheless, 'svelte bodies' – which I can't read without thinking of felt and the Burmese cats, Trotter and Otter – is corny, as is the spouting bloodstream and 'shaking with death rattle'.

When I first compared them, I liked the inclusion in the later version of the lines about the children who looked back guiltily when shot, as though having been responsible for breaking 'a vial of expensive perfume'. Now I suspect it would stand out as inauthentic within the 60,000 word collage that I will never, and could never, assemble.

After percussing my spine, my doctor mentioned Tito's silk suits. I said, yes, not having heard of Tito's silk suits. I was lying on a green vinyl bed over which he had placed a disposable sheet. I put my shirt back on and we walked together to his desk. Behind him,

I noticed there were several books with 'hypnotherapy' written on their spines. Several years earlier, before I'd travelled to Sarajevo and had that trouble with the light from the Tower, I had tried hypnotherapy with him to treat my difficulty with sleeping. I'd had one session, after which I briefly changed doctors.

He printed out a referral and told me to go straight to a radiology centre in Balaclava for an X-ray and to book an appointment with him for the following afternoon. At the door, he shook my hand and asked me what a creative writing PhD was. I told him I was writing a collection of stories and that I was trying to work out a suitable form for writing about trauma that I had not personally experienced.

Part 2

In Melbourne, after visiting my doctor, I decided not to go for the X-ray, feeling that I had been overreacting about the slight pain in my back and the occasional loss of sensation and strength in my arms. I stepped back onto the 96 tram, getting off just past Carlton Gardens. Rather than return home immediately, I walked towards Readings Books in Carlton with the intention of buying Alexis Wright's *The Swan Book*. I left Readings Books not with *The Swan Book*, but with the *Collected Stories of Joaquim Maria Machado de Assis*, though I suspected I would never read this either.

Shortly after returning to Melbourne, Jamie and I went to see *A Fifty Year Argument*, a documentary about the *New York Review of Books*. Joan Didion appeared on the screen, thin and frail. Her hair and skin were white. She was wearing a pale peach sweater. A black plastic hairclip stood out from one side of her head like a bow.

Her eyes seemed to want everyone to believe she was bright enough and old enough to understand everything being said. Behind her there were two framed photographs, both out of focus, but clearly of the same young girl, Didion's daughter, Quintana.

Didion disappeared from the screen and Jamie leant toward me and punched me on the shoulder. In our first weeks of going out – before we were certain we *were* going out – Jamie called me to say she had just finished reading *Blue Nights*, Didion's memoir of the death of Quintana. She said that she liked the book, but that she had sobbed through Didion's earlier memoir, *The Year of Magical Thinking*, about the death of her husband John Gregory Dunne. She was disappointed that she hadn't been able to be moved this time around. She said, maybe in some subliminal way, daughters don't matter to me as much as husbands.

The three-way relationship between the civilian dashing across an intersection, the sniper and the photographer features often in writing about the siege of Sarajevo. Two of the three are shooting, the same two who are almost certain to outlive the relationship.

We had arrived in Sarajevo during a heatwave. We'd made an error with our Airbnb booking, so spent our first week in one of the enormous apartment buildings in Dobrinja that were once part of the Olympic Village. I took photographs of the multi-storey Olympic mascot – a fox with an orange scarf – painted on the blank faces at the ends of the blocks, trying to make it obvious to anyone who saw me that I wasn't photographing the patched-up mortar damage.

It was hot in the apartment and when I opened the window on the first night, it swung out, only attached by the bottom hinge. The apartment was on the seventh floor and I couldn't reach the handle of the window but was able to claw it back in by the frame.

This was during Ramadan, so cannons were sounding throughout the day, making us feel less alone.

When Jamie and I were lined up at the Capitol to see *A Fifty Year Argument*, we ran into Paul, who had read at one of the first of the Slow Canoe Readings that I organised. A few days before the reading, his then girlfriend had left him, unexpectedly moving interstate. On the night, he read a surprisingly long story. When the story concluded, or at least ended, he asked, can I read one more? A poet called out, no, that was long enough.

The story had been about whaling in convict-era Australia and sounded as though it was impeccably researched, but I wasn't paying attention to the specifics, being too worried about the restlessness of the audience. I said, I guess so, and Paul read another long and slow story.

Afterwards, as a group of us walked towards Gertrude Street he stopped and hugged me. Then, with his hands on my shoulders, he apologised. His face was so near mine that I could feel the apology travelling from his mouth to mine. I said, you were amazing, and broke away.

At the cinema, he introduced us to his new girlfriend. They were seeing the same film, so ducked under the cable and joined us in the queue. He looked much better than the last time I'd seen him. His hair was white, and he and his girlfriend were both wearing too-small midnight-blue velvet jackets. I asked him how his PhD was going and he said he was going to be submitting in two weeks. He then said, well, the scholarship ran out. If you're anything like me, he said, you'll really start working when the money stops. The title of his PhD had changed a lot. It had always been consistently short: *The Ship*, *The Ocean*, *The Whale*. His title was now *The Maritime Politics of Moby Dick*. Boring, huh, he said with pride.

On our first day in Sarajevo, Jamie and I caught a bus into the city. The day was hot and awful and we regretted going there. We found a bookstore and I bought a copy of Joe Sacco's graphic novel *Safe Area Goražde*. It was about the siege and genocide of Goražde after the UN had, as with Srebrenica, deemed it a 'safe area' for Bosnian Muslims within Serb-held territory, but had provided no military support to enforce it.

I didn't like reading *Safe Area Goražde* but I did read it, sitting on the bed of our apartment in the heat. The kitchen was a narrow passage adjoining the bedroom, which was the size of the bed, and was, in addition, being used to store furniture that didn't otherwise fit in the apartment. We didn't know what to eat and we didn't want to go out. I googled the area, Dobrinja, which I learned was the most heavily bombed part of the city during the siege, being so near the airport.

I watched a YouTube clip about the 'defenders of Dobrinja', showing teenagers and young men with assault rifles who didn't look like they could defend very much. Before coming to Sarajevo, I had bookmarked the *Final Report of the United Nations Commission of Experts Established Pursuant to Security Council Resolution 780*, which included a day-by-day account of the siege. I clicked on it and starting reading. The subheadings of the report for each day were: Combat and Shelling Activity; Targets Hits; Description of Damage; Sniping Activity; Casualties; Narrative of Events.

Description of damage for 26 August 1992:

> Sarajevo's National Library building was still ablaze after Tuesday's shelling; the shelling of the [library] reportedly touched off fires which destroyed many of the 3,000,000 volumes contained within the building.

Jamie told me to stop – it was too hot and she didn't want to hear anything more. We watched the third season of *Game of Thrones* on her laptop. When we were done, we went back and watched the first two. I thought, this isn't so different from *Safe Area Goražde*, but it was better.

My relationship with my supervisor at this time was not going well. In the weeks leading up to my departure, she'd made it clear that she no longer thought the trip was a good idea. She said, why don't you go up north for a while instead?

From Sarajevo, I spoke to my co-supervisor, who I liked very much, via Skype. She held up a marked-up printout of an essay of mine I'd emailed her on Imre Kertész's *Detective Story* and Jenny Erpenbeck's *The Book of Words*. She swished it around in front of the camera. She said, with exasperation, what are you doing? It's all quotes. You don't *say* anything.

I received an email from my supervisor linking to an article titled 'The Holocaust Just Got More Shocking' in the *New York Times*. The article was about a 13-year-long mapping project by the United States Holocaust Memorial Museum of all the ghettos, slave labour camps, concentration camps, and death camps around Europe during the Second World War. In total, they'd found 42,500 facilities. My supervisor wrote, people *always* know. At the bottom of her email she wrote, I don't know what you're doing over there with all your time. She said, if she didn't see some work from me soon I would have to either reconsider continuing my candidature or at least returning home.

Early one morning in Sarajevo, after we'd moved near the twist tower, when I couldn't sleep, I took my camera and left the apartment. It was light enough to see and I wondered why the tower was still lit up. I walked over the back of our hill, down a great many steps at the bottom of which there was a market. I bought a bag of subtly and deliciously fermented figs from a man at a card table. When I handed him 20 convertible marks, he pulled a clear plastic shopping bag of change from his jacket pocket.

On the other side of the market I bought a grey baseball cap because it was drizzling and I didn't want wet hair in the cold. Wearing the grey baseball cap along with my navy-blue puffer jacket, I felt like a real photographer.

From the market, I walked up a hill toward the 1984 Winter Olympic facilities. I cut through a line of trees to a smashed-up staircase leading up to the speed skating long track and climbed over a low wooden fence. I stood against a tall metal gate. I'd read that the track hadn't had ice on it since the war. It was concrete. The bleachers that must have once been here were gone. At one end stood a tall square concrete column with the Olympic rings along the top with the orange Sarajevo Olympic Games symbol below. The symbol could either look like a sun or four u-shaped office workstations pushed together. There was a cemetery to my right.

In the photographs and footage I'd seen of the 1984 Winter Olympics, the track and the neighbouring sports centre, Zetra – the location of the indoor rink for short track and figure skating – were surrounded by parkland. The area had now been filled with white Islamic gravestones crudely shaped like people.

I couldn't imagine anything better than racing around an open-air 200-metre-long track in this valley in a good pair of speed skates. I imagined doing so with my brother, Daniel, leading a lap each and then stepping out, tucking behind the other to be pulled along in the draft. I could hear the clean silence of blades on ice.

Standing on the concreted ice-skating rink, I took black-and-white photographs of the track through the gate. They were just the kind of moody photographs I liked. I walked down the steps and around the speed skating long track toward Zetra. Zetra had been bombed early during the siege. What was left of it was used as a morgue, which I suppose explains why this park is a cemetery now. It was later rebuilt when it was found the footings were intact.

Further on, beyond the main stadium, there was a tartan running track. I couldn't see anyone, so I walked towards and then around the track. It had six lanes on the main straight but only three on the bends and on the back straight. There were houses right up against the back straight with windows looking onto the track. The tartan had been patched with something blue and softer than the tartan, and something black and harder than the tartan.

I didn't know if the holes were from wear, or from mortar. At one end in the centre of the track there were two red-clay tennis courts. At the other end, running across, there was a short soccer field. The goal nearest the home straight was tipped over. The long-jump pit was covered with new black plywood.

Above the track, there was an undulating running circuit that looped in a figure eight around two hammer throwing fields. I walked up the steep slope and around the figure eight. On one side, there was a slim woman walking back and forth on the grass carrying a throwing hammer.

I checked the time on my phone and realised I'd stayed out longer than I'd intended. I hadn't noticed how bright the morning had become and that the road I could see going up the hill now had cars going down it regularly. I walked back towards the apartment, around the main stadium. I stopped to look at what looked like a 1980s sign painted on the side of the stadium, though it was in English, and then noticed a litter of puppies scattered around a shrub below the sign. I walked closer and saw there were an overturned box and an overturned bucket that would have contained water. They were too young to look after themselves.

The puppies climbed over my shoes, looking up at me as though I was the new centre of their universe. When I took Jamie to see them later that day they climbed over her shoes too. She said, I wish you hadn't brought me here.

In our final few weeks in the city, along with cold, the smog set in. Everywhere smelt like burning tyres. One morning after a few days, desperate to wake up and desperate to sleep, I jogged down to the tartan running track. I was too tired to run many laps, so I sprinted the straights and jogged the bends. I was the only one there. Every time I'd been down to the track, since the first time, I had found people running, and people playing tennis and playing soccer. I felt sick from breathing in the smell of burnt tyres but I kept running.

A few days after seeing *A Fifty Year Argument* I ran into Paul in the library off Flinders Lane. The film festival was still going and he was on his way to see a film about dwarves breaking into a nuclear power plant. We walked towards Swanston Street together. He stopped at Lord of the Fries and said, I'm just going to get some fries. I said that Jamie and I were going down the coast for a few weeks and then we should all have dinner. He said, you've got my digits, right? He was wearing the blue velvet jacket again. No, I didn't think I did, I said, reaching for my phone. But I couldn't think of his full name. I could only think of 'Paul'.

During a later meeting with my supervisor, I mentioned that I'd been reading and thinking about silence and about Aboriginal violence. We were in her office. She crossed the room and closed the door and said in a hushed voice, violence *by* Aboriginals, or *on* Aboriginals? I said *on*, and she looked relieved.

Later in the meeting, I told her that I'd grown up in Cairns. She said that I must therefore have had interesting personal experiences relevant to our discussions. I said yes in a way that I hoped conveyed a dark pooling of insight. I knew I should not talk about the few and meagre experiences I'd had, but I did.

I told her about the man who had appeared at our door in the Little Street house with blood on his face and clothing. I'm not certain I saw the man, however clearly I see him now. I am relatively certain, however, that I saw his blood on and around the front passenger seat of our white Mitsubishi Econovan the following day from when my father had driven him to the Base Hospital on the Esplanade.

My supervisor was listening and not responding. I then told her about being 11 or 12 and running out of a video game arcade with my friends. I was running along the paved footpath of Lake Street when a man grabbed my arm and then my shirt and thumped me against a glass shop front. He said, if I see another one of you … I thought he must have been referring to me and my friends' long hair and oversized t-shirts. He let go of me and said he was sorry, and that I was to ignore him. My friends kept running, and when I soon found them around the next street corner, one of them pulled the air hockey puck that we had been using in the arcade out of his pocket and said, 'look'. I thought, without the table, it's just a plastic disc.

While telling this story to my supervisor, I realised I had no idea if the man was Aboriginal, or why I'd ever thought he was. I knew I should stop now, that I'd made an error.

I then told her about Zach, who definitely wasn't Aboriginal. He was Islander, and that was close. One afternoon after school, my year five teacher, Miss Brooks, asked to talk to me. I was very fond of Miss Brooks because she noticed I was fast with times tables, because she told me my freckles would fade, and because on sports days she wore perfect white and pinky-orange Nike Air 180s.

Miss Brooks told me that she knew I had called something out to Zach that I shouldn't have. We'd been playing soccer on the edge of the grass running track, and the bell to signal the end of recess had sounded, but Zach wouldn't pass the ball to me one last time. I called out, 'Pass it before you melt.' She explained that I mustn't say that kind of thing and I felt crushed that she knew I had.

At camp that year, Zach started running around the grounds as though to escape. Something had happened that morning or there was talk of something happening during the night. As far as we all knew, no one had ever outrun Zach. We didn't expect him to be caught, but if he were to be, each one of us wanted to be the one that did it. I recall a line of children in the bright over-sized clothing of the time – large blocks of colour – along with teachers, chasing him around the pine trees and blue dormitories of the Genazano Centre.

Some time after moving to the better apartment in Sarajevo, we visited the Twist Tower, having read on Trip Adviser about the view from the cafe. We walked down the hill, past an apartment building three down from ours, which, like the rest of the apartment buildings on the street, looked like a small ski hotel. Four teenage girls would often be sitting on the stairs there.

When the girls saw us that day, they sung in unison, but with no enthusiasm, *I guess what you say is true / I could never be the right kind of girl for you / I could never be your woman.* One of them called something out that we couldn't understand.

We didn't discuss it, but my understanding is that Jamie and I always tried to be like our idea of ideal parents whenever we passed them: calm and yet not susceptible to taking shit. We failed each time and could only fail; they wanted to be seen and heard, but not acknowledged.

We turned right off the road onto a staircase slicing down to the same winding road below. Sections of hand railing had been cut away. Later, when I would run up and down these stairs from time to time, I would think of the homemade steel-piping rifles in the Sarajevo Museum. I thought, how dumb and feeble it was to make the connection.

Further along, the stairs gave way to a gravel path with long grass on either side. On our left there was a concrete frame of something the size of a house – ruins or a building that was never completed. In the grass, nesting, I suddenly noticed a straw-coloured dog with a helmet of hard blood. I raced forward, pulling Jamie along. She didn't see what had frightened me.

In the Twist Tower cafe, we ordered and paid for two coffees at the counter and walked around the big curved windows, looking down at the view. We found the river, the old town, Baščaršija, the National Library which had almost been rebuilt, but which would now be the town hall rather than a library. A man appeared through double swing doors with our coffees and left them at the table where our coats were.

We could see all the way out to the enormous patched-up residential towers in Dobrinja, where we stayed for the first week, during the heat wave. We discovered a large forested area behind our apartment, spreading over the crest of the hill. It must be where the howling came from at night, Jamie said.

I thought about the scalp of the dog nesting in the grass and Ted Hughes's description of a jaguar's head as the worn-down stump of another whole jaguar. Sarajevo, I thought, was a worn-down stump of another whole Balkans. I said this to Jamie, who'd sat down in front of one of the coffees, but aloud it sounded showy and imprecise. Jamie said, the line was wasted on a poem about an animal.

Disclosure statement

No potential conflict of interest was reported by the author.

References

Hemon, Aleksandar. "A Coin." *Chicago Review* 43.1 (1997): 61–74.

Hemon, Aleksandar. "A Coin." *The Question of Bruno*. New York: Vintage, 2001. 117–34.

Malcolm, Janet. "Art of fiction No. 4: interview with Katie Roiphe." *Paris Review* 196 (2011): 126–51.

Mehmedinović, Semezdin. "Soviet Computer." Trans. Una Tanović. *Massachusetts Review* 52.3&4 (2011): 416–32.

Weine, Stevan. *Testimony after Catastrophe: Narrating the Traumas of Political Violence*. Evanston, Illinois: Northwestern University Press, 2008.

Index